The Cambridge Manuals of Science and Literature

THE STORY OF A LOAF OF BREAD

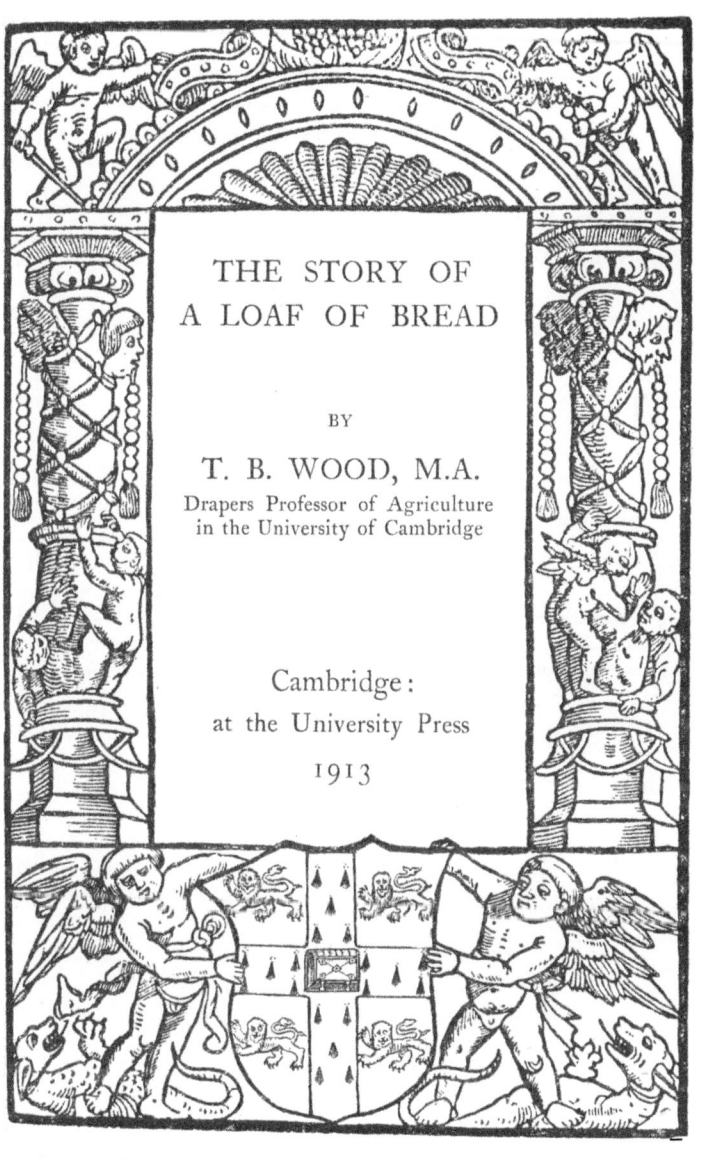

THE STORY OF A LOAF OF BREAD

BY

T. B. WOOD, M.A.

Drapers Professor of Agriculture
in the University of Cambridge

Cambridge:
at the University Press
1913

CAMBRIDGE UNIVERSITY PRESS
Cambridge, New York, Melbourne, Madrid, Cape Town,
Singapore, São Paulo, Delhi, Tokyo, Mexico City

Cambridge University Press
The Edinburgh Building, Cambridge CB2 8RU, UK

Published in the United States of America by Cambridge University Press, New York

www.cambridge.org
Information on this title: www.cambridge.org/9781107606067

© Cambridge University Press 1913

First published 1913
First paperback edition 2011

A catalogue record for this publication is available from the British library

ISBN 978-1-107-60606-7 Paperback

Cambridge University Press has no responsibility for the persistence or
accuracy of URLs for external or third-party internet websites referred to in
this publication, and does not guarantee that any content on such websites is,
or will remain, accurate or appropriate.

*With the exception of the coat of arms
at the foot, the design on the title page is a
reproduction of one used by the earliest known
Cambridge printer, John Siberch,* 1521

PREFACE

I HAVE ventured to write this little book with some diffidence, for it deals with farming, milling and baking, subjects on which everyone has his own opinion. In the earlier chapters I have tried to give a brief sketch of the growing and marketing of wheat. If I have succeeded, the reader will realise that the farmer's share in the production of the staple food of the people is by no means the simple affair it appears to be. The various operations of farming are so closely interdependent that even the most complex book-keeping may fail to disentangle the accounts so as to decide with certainty whether or not any innovation is profitable. The farmer, especially the small farmer, spends his days in the open air, and does not feel inclined to indulge in analytical book-keeping in the evening. Consequently, the onus of demonstrating the economy of suggested innovations in practice lies with those who make the suggestions. This is one of the many difficulties which confronts everyone who sets out to improve agriculture.

In the third and fourth chapters I have discussed the quality of wheat. I have tried to describe the investigations which are in progress with the object of improving wheat from the point of view of both the farmer and the miller, and to give some account of the success with which they have been attended. Incidentally I have pointed out the difficulties which pursue any investigation which involves the cultivation on the large scale of such a crop as wheat, and the consequent need of adopting due precautions to ensure accuracy before making recommendations to the farmer. Advice based on insufficient evidence is more than likely to be misleading. Every piece of misleading advice is a definite handicap to the progress of agricultural science.

The fifth chapter is devoted to a short outline of the milling industry. In chapter VI the process of baking is described. In the last two chapters the composition of bread is discussed at some length. I have tried to state definitely and without bias which points in this much debated subject are known with some certainty, and which points require further investigation.

Throughout the following pages, but especially in chapters III and IV, I have drawn freely upon the work of my colleagues. I am also much indebted to my friends, Mr A. E. Humphries, the chairman of the Home Grown Wheat Committee, and Mr E. S. Beaven of Warminster, whose advice has always been at my disposal. A list of publications on the various branches of the subject will be found at the end of the volume.

T. B. W.

GONVILLE AND CAIUS COLLEGE,
    CAMBRIDGE.
        3 *December*, 1912.

CONTENTS

CHAP.		PAGE
	Preface	v
I.	Wheat-growing	1
II.	Marketing	15
III.	The quality of wheat	27
IV.	The quality of wheat from the miller's point of view	51
V.	The milling of wheat	74
VI.	Baking	91
VII.	The composition of bread	108
VIII.	Concerning different kinds of bread	120
	Bibliography	136
	Index	139

LIST OF ILLUSTRATIONS

FIG.		PAGE
1.	Typical ears of wheat	30
2.	Bird-proof enclosure for variety testing	34
3.	A wheat flower to illustrate the method of cross-fertilising	41
4.	Parental types and first and second generation	43
5.	Parent varieties in bird-proof enclosure	48
6.	Testing new varieties in the field	50
7.	Loaves made from Manitoba wheat	54
8.	Loaves made from English wheat	54
9.	Loaves made from Rivet wheat	55
10.	Loaves made from Manitoba wheat, English wheat, and Manitoba-English hybrid, Burgoyne's Fife	59
11.	Gluten in water and acid	69
12.	Gluten in water containing both acid and salts	71
13.	End view of break rolls	81
14.	Break rolls showing gearing	82
15.	Reduction rolls	87
16.	Baking test: loaves rising in incubator	92
17.	Baking test: loaves leaving the oven	93

THE STORY OF A LOAF OF BREAD

CHAPTER I

WHEAT GROWING

WHEAT is one of the most adaptable of plants. It will grow on almost any kind of soil, and in almost any temperate climate. But the question which concerns the wheat grower is not whether he can grow wheat, but whether he can grow it profitably. This is a question of course that can never receive a final answer. Any increase in the price of wheat, or any improvement that lowers the cost of cultivation, may enable growers who cannot succeed under present conditions to grow wheat at a profit. Thus if the population of the world increases, and wheat becomes scarce, the wheat-growing area will doubtless be extended to districts where wheat cannot be grown profitably under present conditions. A study of the history of wheat-growing in this country during the last century shows that the reverse of this took place. In the first half of that period the population had increased, and from lack of transport facilities and other causes the importation of foreign wheat was small. Prices were high in consequence and every acre of available land was under wheat. As transport facilities increased wheat-growing areas were developed in Canada, in the Western States of America, in the Argentine, and in Australia, and the importation of foreign wheat increased enormously. This led to a rapid decrease in prices, and wheat-growing had to be abandoned on all but the most suitable soils in the British Isles. From 1880 onwards thousands of acres of land which had grown wheat profitably for many years were laid down to grass. In the last decade the world's population has increased faster than the wheat-growing area has been extended. Prices have consequently risen, and the area under wheat in the British Isles will no doubt increase.

But although it cannot be stated with finality on what land wheat can be grown, or cannot be grown, at a profit, nevertheless accumulated experience has shown that wheat grows best on the heavier kinds of loam soils where the rainfall is between 20 and 30 inches per annum. It grows nearly as well on clay soils and on lighter loams, and with the methods of dry farming followed in the arid regions of the Western States and Canada, it will succeed with less than its normal amount of rainfall.

It is now about a hundred years since chemistry was applied with any approach to exactitude to questions affecting agriculture; since for instance it was first definitely recognised that plants must obtain from their surroundings the carbon, hydrogen, oxygen, nitrogen, phosphorus, sulphur, potassium, calcium, and other elements of which their substance is composed. For many years there was naturally much uncertainty as to the source from which these several elements were derived. Experiment soon showed that carbon was undoubtedly taken from the air, and that its source was the carbon dioxide poured into the air by fires and by the breathing of animals. It soon became obvious too that plants obtain from the soil water and inorganic salts containing phosphorus, sulphur, potassium, calcium, and so on; but for a long time the source of the plants' supply of nitrogen was not definitely decided. Four-fifths of the air was known to be nitrogen. The soil was known to contain a small percentage of that element, which however amounts to four or five tons per acre. Which was the source of the plants' nitrogen could be decided only by careful experiment. As late as 1840 Liebig, perhaps the greatest chemist of his day, wrote a book on the application of chemistry to agriculture. In it he stated that plants could obtain from the air all the nitrogen they required, and that, to produce a full crop, it was only necessary to ensure that the soil should provide a sufficient supply of the mineral elements, as he called them, phosphorus, potassium, calcium, etc. Now of all the elements which the farmer has to buy for application to his land as manure, nitrogen is the most costly. At the present time nitrogen in manures costs sevenpence per pound, whilst a pound of phosphorus in manures can be bought for fivepence, and a pound of potassium for twopence. The importance of deciding whether it is necessary to use nitrogen in manures needs no further comment. It was to settle definitely questions like this that John Bennet Lawes began his experiments at his home at Rothamsted, near Harpenden in Hertfordshire, on the manuring of crops. These experiments were started almost simultaneously with the publication of Liebig's book, and many of Lawes' original plots laid out over 70 years ago are still in existence. The results which he obtained in collaboration with his scientific colleague, Joseph Henry Gilbert, soon overthrew Liebig's mineral theory of manuring, and showed that in order to grow full crops of wheat it is above all things necessary to ensure that the soil should be able to supply plenty of nitrogen. Thus it was found that the soil of the Rothamsted Experiment Station was capable of growing wheat continuously year after year. With no manure the average crop was only about 13 bushels per acre. The addition of a complete mineral manure containing phosphorus, calcium, potassium, in fact all the plant wants from the soil except nitrogen, only increased the crop to 15 bushels per acre. Manuring with nitrogen on the other hand increased the crop to 21 bushels per acre. Obviously on the Rothamsted soil wheat has great difficulty in getting all the nitrogen it wants, but is well able to fend for itself as regards what Liebig called minerals. This kind of experiment has been repeated on almost every kind of soil in the United Kingdom, and it is found that the inability of wheat to supply itself with nitrogen applies to all soils, except the black soils of the Fens which contain about ten times more nitrogen than the ordinary arable soils of the country. It is the richness in nitrogen of the virgin soils of the Western States and Canada, and of the black soils of Russia, that forms one of the chief factors in their success as wheat-growing lands. It must be added, however, that continuous cropping without manure must in time exhaust the stores of nitrogen in even the richest soil, and when this time comes the farmers in these at present favoured regions will undoubtedly find wheat-growing more costly by whatever sum per acre they may find it necessary to expend in nitrogenous manure. The world's demand for nitrogenous manure is therefore certain to increase. Such considerations as these inspired Sir William Crookes' Presidential address to the British Association in 1898, in which he foretold the probability of a nitrogen famine, and explained how it must lead to a shortage in the world's wheat supply. The remedy he suggested was the utilization of water-power to provide the energy for generating electricity, by means of which the free nitrogen of the air should be brought into combination in such forms that it could be used for manure. It is interesting to note that these suggestions have been put into practice. In Norway, in Germany, and in America waterfalls have been made to drive dynamos, and the electricity thus generated has been used to make two new nitrogenous manures, calcium nitrate and calcium cyanamide, which are now coming on to the market at prices which will compete with sulphate of ammonia from the gas works, nitrate of soda from Chili, Peruvian guano, and the various plant and animal refuse materials which have up to the present supplied the farmer with his nitrogenous manures. This is welcome news to the wheat grower, for the price of manurial nitrogen has steadily risen during the last decade.

Before leaving the question of manuring one more point from the Rothamsted experiments must be referred to. It has already been mentioned that when manured with nitrogen alone the Rothamsted soil produced 21 bushels of wheat per acre. When, however, a complete manure containing both nitrogen and minerals was used the crop rose to 35 bushels per acre which is about the average yield per acre of wheat in England. This shows that although the yield of wheat is dependent in the first place on the nitrogen supplied by the soil, it is still far from independent of a proper supply of minerals. A further experiment on this point showed that minerals are not used up by the crop to which they are applied, and that any excess left over remains in the soil for next year. This is not the case with nitrogenous manures. Whatever is left over from one crop is washed out of the soil by the winter rains, and lost. Translated into farm practice these results mean that nitrogenous manures should be applied direct to the wheat crop, but that wheat may as a rule be trusted to get all the minerals it wants from the phosphate and potash applied directly to other crops which are specially dependent on an abundant supply of these substances.

At Rothamsted, Lawes and Gilbert adopted the practice of growing wheat continuously on the same land year after year in order to find out as quickly as possible the manurial peculiarities of the crop. This however is not the general system of the British farmer, but it has been carried out with commercial success by Mr Prout of Sawbridgeworth in Hertfordshire. The Sawbridgeworth farm is heavy land on the London clay. Mr Prout's system was to cultivate the land by steam power, to manure on the lines suggested by the Rothamsted experiments, and to sell both grain and straw. Wheat was grown continuously year after year until the soil became infested with weeds, when some kind of root crop was grown to give an opportunity to clean the land. A root crop is not sown until June so that the land is bare for cleaning all the spring and early summer. Such crops also are grown in rows two feet or more apart, and cultural implements can be used between the rows of plants until the latter cover the soil by the end of July or August. After cleaning the land in this way the roots are removed from the land in the winter and used to feed the stock. By this time it is too late to sow wheat, so a barley crop is sown the following spring, and with the barley clover is sown. Clover is an exception to the rule that crops must get their nitrogen from the soil.

On the roots of clover, and other plants of the same botanical order, such as lucerne, sainfoin, beans and peas, many small swellings are to be found. These swellings, or nodules as they are usually called, are produced by bacteria which possess the power of abstracting free nitrogen from the air and transforming it into combined nitrogen in such a form that the clover or other host-plant can feed on it. The clover and the bacteria live in Symbiosis, or in other words in a kind of mutual partnership. The host provides the bacteria with a home and allows them to feed on the sugar and other food substances in its juices, and they in return manufacture nitrogen for the use of the host.

When the clover is cut for hay, its roots are left in the soil, and in them is a large store of nitrogen derived from the air. A clover crop thus enriches the soil in nitrogen and is the best of all preparations for wheat-growing. After the clover, wheat was grown again year after year until it once more became necessary to clean the land. This system of wheat-growing was carried on at Sawbridgeworth for many years with commercial success. It never spread through the country because its success depends on the possibility of finding a remunerative market for the straw. The bulk of straw is so great compared with its price that it cannot profitably be carried to any considerable distance. The only market for straw in quantity is a large town, and there is no considerable area of land suitable for wheat-growing near a sufficiently large town to provide a market for the large output of straw which would result from such a system of farming.

The ordinary practice of the British farmer is to grow his wheat in rotation with other crops. Various rotations are practised to suit the special circumstances of different districts, one might almost say of special farms. This short account of wheat-growing does not profess to give a complete account of even English farming practice. It is only necessary to describe here one rotation in order to give a general idea of the advantages of that form of husbandry.

For this purpose it will suffice to describe the Norfolk or four course rotation. This rotation begins with a root crop, usually Swede turnips, manured with phosphates, and potash too on the lighter lands. This crop, as already described, provides the opportunity of cleaning the land. It produces also a large amount of food for sheep and cattle. Part of the roots are left on the land where they are eaten by sheep during the winter. The roots alone are not suitable for a complete diet. They are supplemented by hay and by some kind of concentrated food rich in nitrogen, usually linseed cake, the residue left when the oil is pressed from linseed. Now an animal only retains in its body about one-tenth of the nitrogen of its diet, so that nine-tenths of the nitrogen of the roots, hay and cake consumed by the sheep find their way back to the land. This practice of feeding sheep on the land therefore acts practically as a liberal nitrogenous manuring. The trampling of the soil in a wet condition in the winter also packs its particles closely together, and increases its water-holding power, in much the same way as the special cultural methods employed in the arid western States under the name of dry farming. The rest of the roots are carted to the homestead for feeding cattle, usually fattening cattle for beef. Again the roots are supplemented by hay, straw, and cake of some kind rich in nitrogen. The straw from former crops is used for litter. Its tubular structure enables it to soak up the excreta of the animals, so that the farmyard manure thus produced retains a large proportion of the nitrogen, and other substances of manurial value, which the animals fail to retain in their bodies. This farmyard manure is kept for future use as will be seen later.

As soon as the sheep have finished eating their share of the turnips they are sold for mutton. It is now too late in the season to sow wheat. The land is ploughed, but the ploughing is only a shallow one, so that the water stored in the deeper layers of the soil which have been solidified by the trampling of the sheep may not be disturbed. The surface soil turned up by the plough is pulverised by harrowing until a fine seed-bed is obtained, and barley is sown early in the spring. Clover and grass seeds are sown amongst the barley, so that they may take firm root whilst the barley is growing and ripening. The barley is harvested in the autumn. The young clover and grasses establish themselves during the autumn and winter, and produce a crop of hay the following summer. This is harvested towards the end of June, and the aftermath forms excellent autumn grazing for the sheep and cattle which are to be fed the next winter.

As soon as harvest is over the farmer hopes for rain to soften the old clover land, or olland as it is called in Norfolk, so that he can plough it for wheat sowing. Whilst he is waiting for rain he takes advantage of the solidity of the soil, produced by the trampling of the stock, to cart on to the olland the farmyard manure produced during the cattle feeding of the last winter. As soon as the rain comes this is ploughed in, and the seed-bed for the wheat prepared as quickly as possible. Wheat should be sown as soon as may be after the end of September, so that the young plant may come up and establish itself, while the soil is yet warm from the summer sun, and before the winter frosts set in. The wheat spends the winter in root development, and does not make much show above ground until the spring. It is harvested usually some time in August. The wheat stubble is ploughed in the autumn and again in the spring, and between then and June, when the roots are sown, it undergoes a thorough cleaning.

The complete rotation has now been described. It remains only to point out some of its numerous advantages. In the first place the system described provides excellent conditions for growing both wheat and barley in districts where the rainfall is inclined to be deficient, say from 20 to 25 inches per annum, as it is in the eastern counties, and on the Yorkshire wolds. Not only is an abundant supply of nitrogen provided for these crops through the medium of the cake purchased for the stock, but the solidification of the deeper layers of the soil ensures the retention of the winter's rain for the use of the crop during the dry summer. The residue of the phosphates and potash applied to the root crop, and left in the soil when that crop is removed, provides for the mineral requirements of the barley and the wheat. Thus each crop gets a direct application of the kind of manure it most needs. Rotation husbandry also distributes the labour of the farm over the year. After harvest the farmyard manure is carted on to the land. This is followed by wheat sowing. In the winter there is the stock to be fed. The spring brings barley sowing, the early summer the cleaning of the land for the roots. Then follow the hay harvest and the hoeing of the roots, and by this time corn-harvest comes round once again.

It must not be forgotten that each crop the farmer grows is subject to its own pests. On a four course rotation each crop comes on the same field only once in four years. Whilst the field is under roots, barley, and clover, the wheat pests are more or less starved for want of food, and their virulence is thereby greatly diminished. The catalogue of the advantages of rotation of crops is a long one but one more must be mentioned. The variety of products turned out for sale by the rotation farmer ensures him against the danger which pursues the man who puts all his eggs in one basket. The four course farmer produces not only wheat and barley, but beef and mutton. The fluctuations in price of these products tend to compensate each other. When corn is cheap, meat may be dear, and vice versâ. Thus in the years about 1900, when corn was making very low prices, sheep sold well, and the profit on sheep-feeding enabled many four course farmers to weather the bad times.

The system of wheat-growing above described is an intensive one. The cultivation is thorough, the soil is kept in good condition by manuring, or by the use of purchased feeding stuffs, and the cost of production is comparatively high. Such systems of intensive culture prevail in the more densely populated countries, but the bulk of the world's wheat supply is grown in thinly populated countries, where the methods of cultivation are extensive. Wheat is sown year after year on the same land, no manure is used, and tillage is reduced to a minimum. This style of cultivation gradually exhausts the fertility of the richest virgin soil, and its cropping capacity falls off. As soon as the crop falls below a certain level it ceases to be profitable. No doubt the fertility of the exhausted soil could be restored by suitable cultivation and manuring, but it is usually the custom to move towards districts which are still unsettled, and to take up more virgin soil. Thus the centre of the area of wheat production in the States has moved nearly 700 miles westward in the last 50 years.

CHAPTER II

MARKETING

In the last chapter we have followed the growing of the wheat from seed time to harvest. But when the farmer has harvested his corn his troubles are by no means over. He has still to thrash it, dress it, sell it, and deliver it to the mill or to the railway station. In the good old times a hundred years ago thrashing was done by the flail, and found work during the winter for many skilled labourers. This time-consuming method has long disappeared. In this country all the corn is now thrashed by machines, driven as a rule by steam, but still in some places by horse-gearing. The thrashing machine, like all other labour saving devices, when first introduced was bitterly opposed by the labourers, who feared that they might lose their winter occupation and the wages it brought them. In the life of Coke of Norfolk, the first Lord Leicester, there is a graphic account of the riots which took place when the first thrashing machine was brought into that county.

Only the larger farmers possess their own machines. The thrashing on the smaller farms is done by machines belonging to firms of engineers, which travel the country, each with its own team of men. These machines will thrash out more than 100 bags of wheat or barley in a working day. The more modern machines dress the corn so that it is ready for sale without further treatment. After it is thrashed the wheat is carried in sacks into the barn and poured on to the barn floor. It is next winnowed or dressed, again by a machine, which subjects it to a process of sifting and blowing in order to remove chaff, weed-seeds and dirt. As it comes from the dressing machine it is measured into bags, each of which is weighed and made up to a standard weight ready for delivery. In the meantime the farmer has taken a sample of the wheat to market. The selling of wheat takes place on market day in the corn hall, or exchange, with which each market town of any importance is provided. In the hall each corn merchant in the district rents a small table or desk, at which he stands during the hour of the market. The farmer takes his sample from one merchant to another and sells it to the man who offers him the highest price. The merchant keeps the sample and the farmer must deliver wheat of like quality. In the western counties it is sometimes customary for the farmers to take their stand near their sample bags of corn whilst the merchants walk round and make their bids.

But unfortunately it too often happens that the struggling farmer cannot have a free hand in marketing his corn. In many cases he must sell at once after harvest to raise the necessary cash to buy stock for the winter's feeding. This causes a glut of wheat on the market in the early autumn, and the price at once drops. In other cases the farmer has bought on credit last winter's feeding stuffs, or last spring's manures, and is bound to sell his wheat to the merchant in whose debt he finds himself, and to take the best price offered in a non-competitive market.

These are by no means all the handicaps of the farmer who would market his corn to the best advantage. Even the man who is blessed with plenty of ready money, and can abide his own time for selling his wheat, is hampered by the cumbrous weights and measures in use in this country, and above all by their lack of uniformity. In East Anglia wheat is sold by the coomb of four bushels. By common acceptance however the coomb has ceased to be four measured bushels, and is always taken to mean 18 stones or $2\frac{1}{4}$ cwt. This custom is based on the fact that a bushel of wheat weighs on the average 63 pounds, and four times 63 pounds makes 18 stones. But this custom is quite local. In other districts the unit of measure for the sale of wheat is the load, which in Yorkshire means three bushels, in Oxfordshire and Gloucestershire 40 bushels, and in parts of Lancashire 144 quarts. Another unit is the boll, which varies from three bushels in the Durham district to six bushels at Berwick. It will

18 THE STORY OF A LOAF OF BREAD [CH.

be noted that most of the common units are multiples of the bushel, and it might be imagined that this would make their mutual relations easy to calculate. This however is not so, for in some cases it is still customary to regard a bushel as a measure of volume and to disregard the variation in weight. In other cases the bushel, as in East Anglia, means so many pounds, but unfortunately not always the same number. Thus the East Anglian bushel is 63 pounds, the London bushel on Mark Lane Market is the same, the Birmingham bushel is only 62 pounds, the Liverpool and Manchester bushel 70 pounds, the Salop bushel 75 pounds, and in South Wales the bushel is 80 pounds. Finally, wheat is sold in Ireland by the barrel of 280 pounds, on Mark Lane by the quarter of eight bushels of 63 pounds, imported wheat in Liverpool and Manchester by the cental of 100 pounds, and the official market returns issued by the Board of Agriculture are made in bushels of 60 pounds. There is, however, a growing tendency to adopt throughout the country the 63 pound bushel or some multiple thereof, for example the coomb or quarter, as the general unit, and the use of the old-fashioned measures is fast disappearing.

The farmer of course knows the weights and measures in use in his own and neighbouring markets, but unless he takes the trouble to look up in a book of reference the unit by which wheat is sold at other markets, and to make a calculation from that unit into the unit in which he is accustomed to sell, the market quotations in the newspapers are of little use to him in enabling him to follow the fluctuations of the price of wheat. Thus a Norfolk farmer who wishes to interpret the information that the price of the grade of wheat known as No. 4 Manitoba on the Liverpool market is 7/3½, must first ascertain that wheat is sold at Liverpool by the cental of 100 pounds. To convert the Liverpool price into price per coomb, the unit in which he is accustomed to sell, he must multiply the price per cental by 252, the number of pounds in a coomb of wheat, and divide the result by 100, the number of pounds in a cental; thus:

$$7/3\tfrac{1}{2} \times 252 \div 100 = 18/4\tfrac{1}{2}.$$

It is evident that the farmer who wishes to follow wheat prices in order to catch the best market for his wheat, must acquaint himself with an extremely complicated system of weights and measures, and continually make troublesome calculations. The average English farmer is an excellent craftsman. He is unsurpassed, indeed one may safely say unequalled, as a cultivator of the land, as a grower of crops, and as a breeder and feeder of stock, but like most people who lead open-air lives, he is not addicted to spending his evenings in arithmetical calculations. The corn merchant, whose business it is to attend to such matters, is therefore at a distinct advantage, and the farmer loses the benefit of a rise in the market until the information slowly filters through to him. No doubt the time will come, when not only wheat selling, but all business in this country, will be simplified by the compulsory enactment of sale by uniform weight. The change from the present haphazard system or want of system would no doubt cause considerable temporary dislocation of business, and would abolish many ancient weights and measures, interesting to the historian and the archaeologist in their relations to ancient customs, but in the long run it could not but expedite business, and remove one of the many handicaps attaching to the isolated position of the farmer.

Having sold his wheat the farmer now puts it up in sacks of the standard of weight or measure prevailing in his district. If the merchant who bought it happens to be also a miller, as is frequently the case, the wheat is delivered to the mill. Otherwise it is sent to the railway station to the order of the merchant who bought it. Meantime the merchant has probably sold it to a miller in a neighbouring large town, to whom he directs the railway company to forward it. Thus the wheat directly or indirectly finds its way to a mill, where it will be mixed with other wheats and ground into flour.

We have now followed wheat production in

England from the ground to the mill. But at the present time home grown wheat can provide only about one-fifth of the bread-stuffs consumed by the population of the United Kingdom, and any account of the growing of wheat cannot be complete without some mention of the methods employed in other countries. The extensive methods of wheat-growing in the more thinly populated countries have already been shortly mentioned. But though their methods of production are of the simplest, the arrangements for marketing their produce are far more advanced in organisation than those already described for the marketing of home grown produce.

For thrashing in Canada and the Western States, travelling machines are commonly used, but they are larger than the machines in use in this country, and the men who travel with them work harder and for longer hours. It is usual for a Canadian travelling "outfit" to thrash 1000 bags of wheat in a day, about ten times as much as is considered a day's thrashing in England. Harvesting and thrashing machinery has evolved to an extraordinary extent in the West on labour saving lines. On the Bonanza farms of the Western States machines are in use which cut off the heads of the wheat, thrash out the seed, and bag it ready for delivery, as they travel round and round the field. Such machines of course leave the straw standing where it grew, and there it is subsequently burnt.

Since wheat is grown every year, few animals are kept beyond the working horses. Very little straw suffices for them and the rest has no value since its great bulk prohibits its profitable carriage to a distance.

After being thrashed the grain is delivered, usually in very large loads drawn by large teams of horses, to the nearest railway station, whence it is despatched to the nearest centre where there is a grain store, or elevator as it is called. Here it is sampled by inspectors under the control, either of the Government or the Board of Trade, as the committee is called which manages the wheat exchange at Chicago or other of the great wheat trading centres. The inspectors examine the sample, and on the result of their examination, assign the wheat to one or other of a definite series of grades. These grades are accurately defined by general agreement of the Board of Trade or by the Government. Each delivery of wheat is kept separate for a certain number of days after it has been graded, in case the owner wishes to appeal against the verdict of the inspector. Such appeals are allowed on the owner forfeiting one dollar per car load of grain if the verdict of the inspector is found to have been correct. At the Chicago wheat exchange 27 grades of wheat are recognised. The following examples show the methods by which they are defined. The definitions are the subject of frequent controversy.

No. 1 Northern Hard Spring Wheat shall be sound, bright, sweet, clean, and shall consist of over 50 per cent. of hard Scotch Fife, and weigh not less than 58 pounds to the measured bushel.

No. 1 Northern Spring Wheat shall be sound, sweet and clean; may consist of hard and soft varieties of spring wheat, but must contain a larger proportion of the hard varieties, and weigh not less than 57 pounds to the measured bushel.

No. 2 Northern Spring Wheat shall be spring wheat not clean enough or sound enough for No. 1, but of good milling quality, and must not weigh less than 56 pounds to the measured bushel.

No. 3 Northern Spring Wheat shall be composed of inferior shrunken spring wheat, and weigh not less than 54 pounds to the measured bushel.

No. 4 Northern Spring Wheat shall include all inferior spring wheat that is badly shrunken or damaged, and shall weigh not less than 49 pounds to the measured bushel.

When sampling wheat for grading, the inspectors also estimate the number of pounds of impurities per bushel, a deduction for which is made under the name of dockage. At the same time the weight of wheat in each car is officially determined. All these points, grade, dockage, and weight, are officially registered, and as soon as the time has elapsed for dealing with any appeal which may arise, the wheat is mixed with all the other wheats of the same grade which may be at the depot, an official receipt for so many bushels of such and such a grade of wheat subject to so much dockage being given to the seller or his agent. These official receipts are as good as cash, and the farmer can realise cash on them at once by paying them into his bank, without waiting for the wheat to be sold.

As each delivery of wheat is graded and weighed, word is sent to the central wheat exchanges that so many bushels of such and such grades are at the elevator, and official samples are also sent on at the same time. The bulk of the sales however are made by grade and not by sample. The actual buying and selling takes place in the wheat exchanges, or wheat pits as they are called, at Chicago, New York, Minneapolis, Duluth, Kansas City, St Louis, and Winnipeg, each of which markets possesses its own special character. Chicago the greatest of the wheat markets of the world passes through its hands every year about 25 million bushels of wheat, chiefly from the western and south-western States. It owes its pre-eminence to the converging railway lines from those States, and to its proximity to Lake Michigan which puts it in touch with water carriage. New York has grown in importance as a wheat market since the opening of the Erie Canal. It is especially the market for export. Minneapolis is above all things a milling centre. No doubt it has become so partly on account of the immense water-power provided by the Falls of St Antony. It receives annually nearly 100 million bushels of wheat, its speciality being the various grades of hard spring wheat. Duluth is the most northern of the American wheat markets. It receives and stores over 50 million bushels annually. It owes its importance to its position on Lake Superior, which is available for water carriage. Kansas City deals with over 40 million bushels per annum, largely hard winter wheat, which it ships down the Missouri River. St Louis deals in soft winter wheats to the extent of about 20 million bushels per annum. Winnipeg is the market for Canadian wheats, to the extent of over 50 million bushels per annum. It has the advantage of two navigable rivers, the Red River and the Assiniboine, and it is also a great railway centre. Its importance is increasing as the centre of the wheat-growing area moves to the north and west, and it is rapidly taking the leading position in the wheat markets of the world.

It has been stated above that Chicago is the greatest wheat market, but it will no doubt have been noticed that this is not borne out by the figures which have been quoted. For instance, Minneapolis receives every year nearly four times as much wheat as Chicago. The reason of this apparent discrepancy is that the sales at Minneapolis are really *bona fide* sales of actual wheat for milling, whilst nine-tenths of the sales at Chicago are not sales of actual wheat, but of what are known as "futures." On this assumption, whilst the actual wheat received at Chicago is 25 million bushels, the sales amount to 250 million bushels. Such dealing in futures takes place to a greater or less extent at all the great wheat markets, but more at Chicago than elsewhere.

The primary reason for dealing in futures is that the merchant who buys a large quantity of wheat, which he intends to sell again at some future time, may be able to insure himself against loss by a fall in price whilst he is holding the wheat he has bought. This he does by selling to a speculative buyer an equal quantity of wheat to be delivered at some future time. If whilst he is holding his wheat prices decline, he will then be able to recoup his loss on the wheat by buying on the market at the reduced price now current to meet his contract with the speculative buyer, and the profit he makes on this transaction will more or less cover his loss on the actual deal in wheat which he has in progress. As a matter of fact he does not actually deliver the wheat sold to the speculative buyer. The transaction is usually completed by the speculator paying to the merchant the difference in value between the price at which the wheat was sold and the price to which it has fallen in the interval. This payment is insured by the speculative buyer depositing a margin of so many cents per bushel at the time when the transaction was made. Speculation is, however, kept within reasonable bounds by the fact that a seller may always be called upon to deliver wheat instead of paying differences.

The advantage claimed for this system of insurance is that whilst it is not more costly to the dealers in actual wheat than any other equally efficient method, it supports a number of speculative buyers and sellers, whose business it is to keep themselves in touch with every phase of the world's wheat supply. The presence of such a body of men whose wits are trained by experience of market movements, and who are ready at any moment to back their judgment by buying and selling large quantities of wheat for future delivery, is considered to exert a steadying effect on the price of wheat, and to lessen the extent of fluctuations in the price.

CHAPTER III

THE QUALITY OF WHEAT

IN discussing the quality of wheat it is necessary to adopt two distinct points of view, that of the farmer and that of the miller. A good wheat from the farmer's point of view is one that will year by year give him a good monetary return per acre. Now the monetary return obviously depends on two factors, the yield per acre and the value per quarter, coomb, or bushel, as the case may be. These two factors are quite independent and must be discussed separately.

We will first confine our attention to the yield per acre. This has already been shown to depend on the presence in the soil of plenty of the various elements required by plants, in the case of wheat nitrogen being especially important. The need of suitable soil and proper cultivation has also been emphasised. These conditions are to a great extent under the control of the farmer, whose fault it is if they are not efficiently arranged. But there are other factors affecting the yield of wheat which cannot be controlled, such for instance as sunshine and rainfall. The variations in these conditions from year to year are little understood, but they are now the subject of accurate study, and already Dr W. N. Shaw, the chief of the Meteorological Office has suggested a periodicity in the yield of wheat, connected with certain climatic conditions, notably the autumnal rainfall.

We have left to the last one of the most important factors which determine the yield of wheat, namely, the choice of the particular variety which is sown. This is undoubtedly one of the most important points in wheat-growing which the farmer has to decide for himself. The British farmer has no equal as a producer of high class stock. He supplies pedigree animals of all kinds to the farmers of all other lands, and he has attained this preeminence by careful attention to the great, indeed the surpassing, importance of purity of breed. It is only in recent years that the idea has dawned on the agricultural community that breed is just as important in plants as in animals. It is extraordinary that such an obvious fact should have been ignored for so long. That it now occupies so prominently the attention of the farmers is due to the work of the agricultural colleges and experiment stations in Sweden, America, and many other countries, and last but by no means least in Great Britain. This demonstration of the value of plant breeding is perhaps the greatest achievement in the domain of agricultural science since the publication of Lawes and Gilbert's classical papers on the manurial requirements of crops.

Wheat is not only one of the most adaptable of plants. It is also one of the most plastic and prone to variation. During the many centuries over which its cultivation has extended it has yielded hundreds of different varieties, whose origin, however, except in a few doubtful cases is unknown. Comparatively few of these varieties are in common use in this country, and even of these it was impossible until recently to

Fig. 1. Typical ears of a few of the many cultivated varieties of wheat

THE QUALITY OF WHEAT

say which was the best. It was even almost impossible to obtain a pure stock of many of the standard varieties. This is by no means the simple matter it appears to be. It is of course quite easy to pick out a single ear, to rub out the grain from it, to sow the grain on a small plot by itself, and to raise a pound or so of perfectly pure seed. This can again be sown by itself, and the produce, thrashed by hand, will give perhaps a bushel of seed which will be quite pure. From this seed it will be possible to sow something like an acre; and now the trouble begins. Any kind of hand thrashing is extremely tedious for the produce of acre plots, and thrashing by machinery becomes imperative. Now a thrashing machine is an extremely complicated piece of apparatus, which it is practically impossible thoroughly to clean. When once seed has been through such a machine it is impossible to guarantee its purity. Contamination in the thrashing machine is usually the cause of the impurity of the stocks of wheat and other cereals throughout the country. The only remedy is for the farmer to renew his stock from time to time from one or other of the seedsmen or institutions who make it their business to keep on hand pure stocks obtained by the method above described.

Comparative trials of pure stocks of many of the standard varieties of wheat, and of the other cereals, are being carried out in almost every county by members of the staff of the agricultural colleges. The object of such trials is to determine the relative cropping power of the different varieties. This might at first sight appear to be an extremely simple matter, but a moment's consideration shows that this is not the case. No soil is so uniform that an experimenter can guarantee that each of the varieties he is trying has the same chance of making a good yield as far as soil is concerned. It is a matter of common knowledge too that every crop of wheat is more or less affected by insect and fungoid pests, whose injuries are unlikely to fall equally on each of the varieties in any variety test. Many other causes of variation, such as unequal distribution of manure, inequalities in previous cropping of the land, irregular damage by birds, may well interfere with the reliability of such field tests.

Much attention has been given to this subject during the last few years, and it has been shown that as often as not two plots of the same variety of wheat grown in the same field under conditions which are made as uniform as possible will differ in yield by 5 per cent. or more. Obviously it is impossible to make comparisons of the cropping power of different varieties of wheat as the result of trials in which single plots of each variety are grown. It is a deplorable fact however that the results of most of the trials which are published are based on single plots only of the varieties compared. Such results can have no claim to reliability. Single plots tests are excellent as local demonstrations, to give the farmers a chance of seeing the general characters of the various wheats in the field, but for the determination of cropping power their results are misleading. For the comparison of two varieties however an accuracy of about 1 per cent., which is good enough for the purpose in view, can be obtained by growing, harvesting and weighing separately, five separate plots of each variety under experiment, provided the plots are distributed in pairs over the experimental field.

Still greater accuracy can be attained by growing very large numbers of very small plots of each variety in a bird-proof enclosure. The illustration shows such an enclosure at Cambridge where five varieties were tested, each on 40 plots. Each plot was one square yard, and the whole 200 plots occupied so small an area that uniformity of soil could be secured by hand culture.

Several experimenters are now at work on these lines, and it is to be hoped that all who wish to carry out variety tests will either follow suit, or content themselves with using their single plots only for demonstrating the general characters of the varieties in the field.

So far we have confined our discussion to the standard varieties, and we must now turn our

Fig. 2. Part of bird-proof enclosure containing many small plots for variety testing

CH. III] THE QUALITY OF WHEAT 35

attention to the work which has been done in recent years on the breeding of new varieties which will yield heavier crops than any of the varieties hitherto in cultivation.

It is impossible to give more than a very brief outline of the vast amount of work which has been done on this subject. Broadly speaking, two methods have been used, selection and hybridisation. Of these selection is the simpler, but even selection is by no means the simple matter it might appear to be. Let us examine for a moment the various characters of a single wheat plant which determine its capacity for yielding grain. The average weight of one grain, the number of grains in an ear, the number of ears on the plant, are obviously all of them characters which will influence the weight of grain yielded by the plant. Many experimenters have examined thousands of plants for these characters, often by means of extremely ingenious mechanical sorting instruments, and have raised strains of seed from the plants showing one or more of these characters in the highest degree. The results of this method of selection have as a rule been unsuccessful, no doubt because the size of the grain, the number of grains in the ear, and the number of ears on the plant, are so largely determined by the food supply, or by some other cause quite outside the plant itself. They are in fact in most cases acquired characters, and are not inherited.

This method of selection results in picking out rather the well nourished plant than the well bred one. Again it is obvious that the weight of grain per acre is measured by the weight of one grain, multiplied by the number of grains per ear, multiplied by the number of ears per plant, multiplied by the number of plants per acre. Selecting for any one of these characters, say large ears, is quite likely to diminish other equally important characters, say number of ears per plant.

In order to avoid these difficulties the method of selection according to progeny has been devised. The essence of this method is to select for stock, not the best individual plant, but the plant whose progeny yields the greatest weight of seed per unit area. This method was applied with great industry and some success in the Minnesota wheat breeding experiments of Willett Hays. Large numbers of promising plants were collected from a plot of the best variety in that district. The seed from each plant was rubbed out and sown separately. One hundred seeds from each plant were sown on small separate plots which were carefully marked out and labelled. Every possible precaution was taken to make all the little plots uniform in every way. By harvesting each plot separately, and weighing the grain it produced, it was possible to find out which of the original plants had given the largest yield. This process

THE QUALITY OF WHEAT

was repeated by sowing again on separate plots a hundred seeds from each individual plant from the best plot, and again weighing the produce of each plot. After several repetitions it was stated that new strains were obtained which yielded considerably greater crops than the variety from which they were originally selected. These results were published in 1895, but no definite statements have since appeared as to the success ultimately attained.

This method of selection is undoubtedly more likely to give successful results than the method which depends on the selection of plants for their apparent good qualities; but it has several weak points. In the first place it is almost impossible to make the soil of a large number of plots so uniform that variation in yield due to varying soil conditions will not mask the variations due to the different cropping power of the seed of the separate plants. Many experimenters are still at work with a view to overcome this difficulty. Secondly, plant breeders are by no means agreed on the exact theoretical meaning of improvement by selection. The balance of evidence at the present time seems to tend towards the general adoption of what is known as the pure-line theory. According to this theory, which was first enunciated by Johannsen of Copenhagen as the outcome of a lengthy series of experiments with beans, the general population of plants, in say a field of wheat of one of the standard varieties giving an average yield of say 40 bushels per acre, consists of a very large number of races each varying in yielding capacity from say 30 to 50 bushels per acre. These races can be separated by collecting a very large number of separate plants, sowing say 100 seeds from each on a separate plot, and weighing the produce separately. The crop on each plot, being the produce of a separate plant, will be a distinct race, or pure line as it is called, and each pure line will possess a definite yielding power of its own. If this is so the difficulty of soil variation can be overcome by saving seed from many of the best plots, and sowing it on several separate plots. At harvest time these are gathered separately and weighed. By averaging the weights of grain from many separate plots scattered over the experimental area the effect of soil variation can be eliminated.

The method is very laborious, but seems to promise successful results. For instance, Beaven of Warminster, working on these lines, has succeeded in isolating a pure line of Archer barley which is a distinct advance on the ordinary stocks of that variety. There appears to be no reason why it should not be applied to wheat with equal success; in fact, Percival of Reading states that his selected Blue Cone wheat was produced in this way. The essence of the method is that if the pure-line theory

THE QUALITY OF WHEAT

holds there is no necessity to continue selecting the best individual plant from each plot, for each plot being the produce of a single plant must be a pure line with its own definite characters. The whole of the seed from a number of the best plots can therefore be saved. The seed from each of these good plots can be used to sow many separate plots: by averaging the yields from these plots the effects of soil variation can be eliminated, and the cropping power thus determined with great accuracy. It is thus possible to pick out the best pure line with far greater certainty than in any other way. It must not be forgotten, however, that the success of the method depends on the truth of the pure-line theory. It should also be pointed out that the cereals are all self-fertilised plants. When working on these lines with plants which are readily cross-fertilised, such for instance as turnips or mangels, it is necessary to enclose the original individual plants, and the subsequent separate plots, so as to prevent them from crossing with plants of other lines, in which case the progeny would be cross-bred and not the progeny of a single plant. This of course enormously increases the difficulty of carrying out the experiment. Enough has been said to show that the task of improving plants by systematic selection is an extremely tedious and difficult one. Of course anyone may be fortunate enough to drop on a valuable sport when carefully inspecting his crops, and it appears likely that many of the most valuable varieties in cultivation have originated from lucky chances of this kind.

It has always been the dream of the plant breeder to make use of the process of hybridisation for creating new varieties, but until the work of Mendel threw new light on the subject the odds were against the success of the breeder. The idea of the older hybridisers was that crossing two dissimilar varieties broke the type and gave rise to greatly increased variation. From the very diverse progeny resulting from the cross, likely individuals were picked out. Seed was saved from these and sown on separate plots, and attempts were made to obtain a fixed type by destroying, or roguing as it is called, all the plants which departed from the desired type. This was a tedious process which seldom resulted in success. Mendel's discoveries, made originally nearly 50 years ago, as the result of experiments in the garden of his monastery, in the crossing of different varieties of garden peas, remained unknown until rediscovered in 1899. In the 12 years which have elapsed since that date the results which have been achieved show clearly that the application of Mendelian methods is likely greatly to increase the simplicity and the certainty of plant improvement by hybridisation.

Perhaps the best way of describing the bearing of Mendel's Laws on the improvement of wheat is to

III] THE QUALITY OF WHEAT 41

give an illustration from the work carried out by Biffen at Cambridge, dealing at first with simple characters obvious to anyone. In one of his first

Fig. 3. A wheat flower with the chaff opened to show the stamens and the stigmas

experiments two varieties of wheat were crossed with each other. The one variety possessed long loose beardless ears, the other short dense bearded ears.

The crossing was performed early in June, sometime before what the farmer calls flowering time. The flowering of wheat as understood by the farmer is the escape of the stamens from the flower. Fertilisation always takes place before this, and crossing must be done of course before self-fertilisation has been effected. The actual crossing is done thus: An ear of one of the varieties having been chosen, one of the flowers is exposed by opening the chaff which encloses it (Fig. 3), the stamens are removed by forceps, and a stamen from a flower of the other variety is inserted, care being taken that it bursts so that the pollen may touch the feathery stigmas. The chaff is then pushed back so that it may protect the flower from injury. The pollen grains grow on the stigmas, and penetrate down the styles into the ovary. In this way cross-fertilisation is effected. It is usual to operate on several flowers on an ear in this way, and to remove the other flowers, so that no mistake may be made as to which seed is the result of the cross. Immediately after the operation the ear is usually tied up in a waxed paper bag. This serves to make it absolutely certain that no other pollen can get access to the stigmas except that which was placed there. At the same time it is a convenient way of marking the ear which was experimented upon. The cross is usually made both ways, each variety being used both as pollen parent and as ovary

III] THE QUALITY OF WHEAT 43

parent. As soon as the cross-fertilised seeds are ripe they are gathered, and early in the autumn they are

Fig. 4. P, P, the two parental types. F_1, the first cross. F_2, 1—6, the types found in the second generation

sown. It is almost necessary to sow them and other small quantities of seed wheat in an enclosure protected by wire netting. Otherwise they are very liable to suffer great damage from sparrows. The plants which grow from the cross-fertilised seeds are known as the first generation. In the case under consideration, they were found to produce ears of medium length and denseness, intermediate between the ears of the two parent varieties, and to be beardless. The first generation plants were also characterised by extraordinary vigour, as is the case with almost all first crosses, both in plants and animals. Their seed was saved and sown on a small plot, and produced some hundreds of plants of the second generation. On examining these second generation plants it was found that the characters of the parent varieties had rearranged themselves in every possible combination, long ears with and without beard, short ears with and without beard, intermediate ears with and without beard, as shown in Fig. 4. These different types were sorted out and counted, when they were found to be present in perfectly definite proportions. This is best shown in the form of a tabulated statement, thus:

Ears	Ears	Ears	Ears	Ears	Ears
Long	Long	Medium	Medium	Short	Short
Beardless	Bearded	Beardless	Bearded	Beardless	Bearded
3	1	6	2	3	1

Translating this into words, out of every 16 plants in the second generation there were four long eared

III] THE QUALITY OF WHEAT 45

plants, three beardless and one bearded; eight plants with ears of intermediate length, six beardless and two bearded; and four short eared plants, three beardless and one bearded. The illustration shows all these types. The experiment has been repeated several times and the same proportions were invariably obtained. The result, too, was independent of the way the cross was made. Seed was collected separately from large numbers of single plants of each type. The seed from each plant was sown by itself in a row, so that its progeny could be separately observed. It was found that all the plants of the second generation possessing ears of intermediate length produced in the third generation plants with long ears, short ears, and medium ears in the proportion of 1 : 1 : 2, the same proportion in fact as in the second generation. Short eared plants produced only short eared offspring, long eared plants only long eared offspring. Bearded plants produced only bearded offspring. Beardless plants, however, produced in some cases only beardless offspring, in other cases both beardless and bearded offspring in the proportion of three of the former to one of the latter. Out of every three beardless plants only one was found to breed true, whilst two gave a mixed progeny. It appears therefore that in the second generation some of the types which occur breed true, whilst others do not. Some of the true breeding individuals can be picked out at sight, for instance, those with long or short bearded ears. Some of those which will not breed true can also be recognised by inspection, for instance, all the plants with ears of intermediate length. In other cases it is only possible to pick out the individual plants which breed true by growing their seed and observing how it behaves. If it produces progeny all of which are like the plant from which the seed was obtained, that plant is a fixed type and will breed true continuously in the future. The final result of the experiment was to obtain in three years from the time the cross was made, four fixed types which subsequent experience has shown breed true continuously, a long eared bearded type, a short eared beardless type, a long eared beardless type and a short eared bearded type. Of these the second two are exactly like the two parental varieties, but the first two are new, each combining one character from each parent. These fixed types already existed in the second generation. Mendel's discoveries with peas showed how to pick them out. Obviously there is no need for the years of roguing by which the older hybridisers used to attempt to fix their desired type. All the types are present in the second generation. Mendel has shown how the fixed ones may be picked out.

The characters described above are not of any great economic importance. Biffen has shown that

III] THE QUALITY OF WHEAT 47

such important characters as baking strength and resistance to the disease known as yellow rust behave on crossing in the same way as beard. Working on the lines of the experiment described above he has succeeded in producing several new varieties which in baking strength and in rust resistance are a distinct advance on any varieties in cultivation in this country. His method of working was to collect wheats from every part of the world, to sow them and to pick out from the crop, which was usually a mixed one, all the pure types he could. These were grown on small plots for several years under close observation. Many were found to be worthless and were soon discarded. Others were observed to possess some one valuable character. Amongst these a pure strain of Red Fife was obtained from Canadian seed, which was found to retain when grown in England the excellent baking strength of the hard wheats of Canada and North America. Again, other varieties were noticed to remain free from yellow rust year after year, even when varieties on adjoining plots were so badly infected that they failed to produce seed. Other varieties, too, were preserved for the sturdiness of their straw, their earliness in ripening, vigour of growth, or yielding capacity. Many crosses were made with these as parents. The illustration shows a corner of the Cambridge wheat-breeding enclosure including a

Fig. 5. Corner of bird-proof enclosure showing a varied assortment of parent varieties of wheat. Crosses have been made on some of them as shown by the ears tied up in paper bags

CH. III] THE QUALITY OF WHEAT 49

miscellaneous collection of parent varieties. The paper bags on the ears show where crosses have been made. From the second generation numbers of individual plants possessing desirable characters were picked out, and the fixed types isolated in the third generation by making cultures from the seed of these single plants. The seed from these fixed types was sown on small field plots, at which stage many had to be rejected because they were found wanting in some character of great practical importance which did not make itself evident in the breeding enclosure. The illustration shows a case in point. It was photographed after heavy rain in July. The weakness of the straw of the variety on the left had not been noticed in the enclosure. The types which approved themselves on the small field plots were again grown on larger plots so that their yield and milling and baking characters could be tested. So far two types have survived the ordeal. One combines the cropping power of the best English varieties with the baking strength of North American hard wheat. It is the outcome of a cross between Rough Chaff and Red Fife. Its average crop in 1911 was 38 bushels per acre as the result of 28 independent trials, and, where the local millers have found out its quality, it makes on the market four or five shillings per quarter more than the ordinary English varieties. The other resulted from a cross between Square

Fig. 6. Field plots of two new varieties of the same parentage which had approved themselves in the bird-proof enclosure. That on the left had to be rejected on account of the weakness of its straw. That on the right is the rust-proof variety known as Little Joss. The photograph was taken after a storm which in the open field found out the weak point of the one variety

Head's Master and a rust-resisting type isolated from a graded Russian wheat called Ghirka. It is practically rust-proof. Consequently it yields a heavier crop than any of the ordinary varieties which are all more or less susceptible to rust. The presence of rust in and on the leaves hinders the growth of the plant, lowers the yield, and increases the proportion of shrivelled grains. It has been estimated that rust diminishes the world's wheat crop by something like one third. The new rust proof variety gave an average yield of about 6 bushels per acre more than ordinary varieties on the average of 28 trials last year. It is called Little Joss and is especially valuable in the Fens and other districts where rust is more than usually virulent.

CHAPTER IV

THE QUALITY OF WHEAT FROM THE MILLER'S POINT OF VIEW

To the miller the quality of wheat depends on three chief factors, the percentage of dirt, weed seeds, and other impurities, the percentage of water in the sample, and a complex and somewhat ill-defined character commonly called strength.

With the methods of growing, cleaning and thrashing wheat practised in Great Britain, practically clean samples are produced, and home grown wheat is therefore on the whole fairly free from impurities. This is, however, far from the case with foreign wheats, many of which arrive at the English ports in an extremely dirty condition. They are purchased by millers subject to a deduction from the price for impurities above the standard percentage which is allowed. The purchase is usually made before the cargo is unloaded. Official samples are taken during the unloading in which the percentage of impurities is determined, and the deduction, if any, estimated.

The percentage of water, the natural moisture as it is usually called, varies greatly in the wheats of different countries. In home grown wheats it is usually 16 per cent., but in very dry seasons it may be much lower, and in wet seasons it may rise to 18 per cent. Foreign wheats are usually considerably drier than home grown wheats. In Russian wheats 12 per cent. is about the average, and that too is about the figure for many of the wheats from Canada, the States, Argentina, and parts of Australia. Indian wheats sometimes contain less than 10 per cent. This is also about the percentage in the wheats of the arid lands on the Pacific coast and in Australia. These figures show that home grown wheats often contain as much as 5 per cent. more water than the foreign

IV] QUALITY OF WHEAT: MILLER'S VIEW

wheats imported from the more arid countries. The more water a wheat contains the less flour it will yield in the mill. Consequently the less its value to the miller. A difference of 5 per cent. of natural moisture means a difference in price of from 1s. 6d. to 2s. per quarter in favour of the drier foreign wheats. This is one of the reasons why foreign wheats command a higher price than those grown in this country.

Turning to the third factor which determines the quality of wheat from the miller's point of view, we may for the present define strength as the capacity for making bread which suits the public taste of the present day. We shall discuss this point more fully when we deal with the baking of bread. At present the only generally accepted method of determining the strength of a sample of wheat is to mill it and bake it, usually into cottage loaves. The strength of the wheat is then determined from their size, shape, texture, and general appearance. A really strong flour makes a large, well risen loaf of uniformly porous texture. Wheats lacking in strength are known as weak. A weak wheat makes a small flat loaf. In order to give a numerical expression to the varying degrees of strength met with in different wheats, the Home Grown Wheat Committee of the National Association of British and Irish Millers have adopted a scale as the result of many thousand milling

54 THE STORY OF A LOAF OF BREAD [CH.

and baking tests. On their scale the strength of the best wheat imported from Canada, graded as No. 1 Manitoban, or from the States graded as No. 1

Fig. 7. Loaves made from No. 1 Manitoba. Strength 100

Hard Spring, is taken as 100, that of the well-known grade of flour known as London Households as 80, and that of the ordinary varieties of home grown

Fig. 8. Loaves made from average English wheat. Strength 65

IV] QUALITY OF WHEAT: MILLER'S VIEW 55

wheat, such as Square Head's Master, Browick, Stand Up, etc., as 65. The strength of most foreign wheats falls within these limits. Thus the strength of Ghirka wheat from Russia is about 85, of Choice White Karachi from India 75, of Plate River wheat from the Argentine 80, etc. The strongest of all wheats is grown in certain districts in Hungary. It is marked above 100 on the scale, but it is not used for bread making. The soft wheats from the more arid regions

Fig. 9. Loaves made from Rivet wheat. Strength 20

in Australia and the States are usually weaker than average home grown samples, and are marked at 60. Rivet or cone wheat, a heavy cropping bearded variety much grown by small holders,—since the sparrow, which would ruin small plots of any other variety, seems to dislike Rivet, possibly on account of its beard,—is the weakest of all wheats, and is only marked at 20, which means that bread baked from Rivet flour alone would be practically unsaleable. Rivet wheat finds a ready sale, however, for making certain kinds of biscuits.

In order to make flour which will bake bread to suit the taste of the general public of the present day, the miller finds it necessary to include in the mixture or blend of wheats which he grinds a certain proportion of strong wheats such as Canadian, American, or Russian. The quantity of strong wheat available is limited. Consequently strong wheat commands a relatively high price. The average difference in price of say No. 1 Manitoban and home grown wheat is about 5*s.* per quarter. It is possible of course that the public taste in bread may change, and damp close textured bread may become fashionable. In this case no doubt the difference in price would disappear. Under present conditions the necessity of including in his grinding mixture a considerable proportion of strong foreign wheat is a distinct handicap against the inland miller as compared with the port miller. The latter gets his foreign wheat direct from the ship in which it is imported, whilst the former has to pay railway carriage from the port to his mill. The question naturally arises—is it not possible to grow strong wheats at home and sell them to the inland miller?

This question has been definitely answered by the work of the Home Grown Wheat Committee

QUALITY OF WHEAT: MILLER'S VIEW

during the last 12 years. The committee collected strong wheats from every country where they are produced, and grew them in England. From the first crop they picked out single plants of every type represented in the mixed produce, for strong wheats as imported are usually grades and not pure varieties. From the single plants they have established pure strains of which they have grown enough to mill and bake. From most of the strong wheats they were unable to find any strain which would produce strong wheat in England. Thus the strong wheat of Hungary when grown in England was no stronger than any of the ordinary typical home grown wheats. But from the strong wheat of Canada was isolated the variety known as Red Fife, which makes up a very large proportion of the higher grades of American and Canadian wheats, and this variety when grown in England was found to continue to produce wheat as strong as the best Canadian. Year after year it has been grown here, and when milled and baked its strength has been found to be 100 or thereabouts on the scale above described. Finally it was found that a strain of Red Fife which had been brought over from Canada 20 years ago, and grown continuously in the western counties ever since, under the name of Cook's Wonder, was still producing wheat which when ground and baked possessed a strength of about 100. Thus it was conclusively proved that in the case of Red Fife at any rate the English climate was capable of producing really strong wheat. The strength of Hungarian and Russian wheats appear to be dependent on the climate of those countries. Red Fife, however, produces strong wheat wherever it is grown. It is interesting to note that this variety although first exploited in Canada and the States is really of European origin. It was taken out to Canada by an enterprising Scotchman called Fife in a mixed sample of Dantzig wheat. He grew it for some time and distributed the seed. Pure strains have from time to time been selected by the American and Canadian experiment stations.

But the discovery that Red Fife would produce strong wheat in England by no means solved the problem, for when the Home Grown Wheat Committee distributed seed of their pure strain of that variety for extended testing throughout the country, it was soon found to be only a poor yielder except in a few districts. A yield of three quarters of strong grain, even if it makes 40s. per quarter on the market, only gives to the farmer a return of £6 per acre, as compared with a return of nearly £8 from $4\frac{1}{2}$ quarters of weak grain worth 35s. per quarter, which can usually be obtained by growing Square Head's Master, or some other standard variety.

It was at this point that Mendel's discoveries came to the rescue. Working on the Mendelian lines

IV] QUALITY OF WHEAT: MILLER'S VIEW 59

already explained, Biffen at Cambridge crossed Red Fife with many of the best English varieties. From one of the crosses he was able to isolate a new variety in which are combined the strength of Red Fife and the vigour and cropping power of the English parent. This variety, known as Burgoyne's Fife, has been grown and distributed by members of the Millers' Association. In 1911 on the average of 28 separate

Fig. 10. The left-hand loaf was made from average English wheat. The loaf in the centre was made from Burgoyne's Fife, and is practically identical in size and shape with the right-hand loaf which was made from imported No. 1 Manitoba

trials it yielded 38 bushels per acre, which is well above the average of the best English varieties. It has been repeatedly milled and baked, and its strength is between 90 and 100, practically the same as that of Red Fife. It has been awarded many prizes at agricultural shows for quality, and it commands on markets where the local millers have found out its baking qualities about the same price as the best foreign strong wheats, that is to say from 4*s.* to 5*s.* per quarter more than the average price of home grown wheat. Taking a fair average yield of wheat as four quarters per acre, Burgoyne's Fife gives to the farmer an increased return over the ordinary varieties of about 16*s.* per acre. The introduction of such a variety makes the production of strong wheat in England a practicable reality, and will be a boon both to the farmer and to the inland miller. It is likely too that the possibility of obtaining a better return per acre will induce farmers to grow more wheat. Anything that tends to increase the production of home grown wheat and makes Great Britain less dependent on foreign supplies is a national asset of the greatest value.

It is of the greatest importance to the miller that he should be able to determine the strength of the wheats he buys. Obviously the method mentioned above, which entails milling enough of the sample to enable him to bake a batch of bread, is far too lengthy to be of use in assessing the value of a sample with a view to purchase. The common practice is for the miller or corn merchant to buy on the reputation of the various grades of wheat, which he confirms by inspection of the sample. Strength is usually associated with certain external characters which can readily be judged by the eye of the practised wheat buyer. Strong wheats are usually red in colour, their

QUALITY OF WHEAT: MILLER'S VIEW

skin is thin and brittle, the grain is usually rather small, and has a very characteristic horny almost translucent appearance. The grains are extremely hard and brittle, and when broken the inside looks flinty. On chewing a few grains the starch is removed and there remains in the mouth a small pellet of gluten, which is tough and elastic like rubber, but not sticky.

Weak wheats as a rule possess none of these characters. Their colour may be either red or white, their skin is commonly thick and tough, the grain is usually large and plump, and often has an opaque mealy appearance. It is soft and breaks easily, and the inside is white, soft and mealy. Very little gluten can be separated from it by chewing, and that little is much less tough and elastic than the gluten of a strong wheat.

These characters, however, are on the whole less reliable than the reputation of the grade of wheat under consideration. To make a reliable estimate of strength from inspection of a sample of wheat requires a natural gift cultivated by continual practice. Even the best commercial judges of wheat have been known to be deceived by a sample of white wheat which subsequent milling and baking tests showed to possess the highest strength. The mistake was no doubt due to the great rarity of strength among white wheats. This rarity will doubtless soon disappear now that a pure strain of White Fife has been isolated and shown to possess strength quite equal to that of Red Fife. Sometimes too the ordinary home grown varieties produce most deceptive samples which show all the external characters of strong wheats. Such samples, however, on milling and baking are invariably found to possess the usual strength of home grown wheat, about 65 on the scale. These considerations show the great need of a scientific method of measuring strength, which can be carried out rapidly and on a small sample of grain. This need is felt at the present time not only by the miller and the merchant, but by the wheat breeder. For instance, in picking out the plants possessing strong grain from cultures of the second generation after making his crosses, the plant breeder up to the present has had to rely on inspection by eye, and on the separation of gluten by chewing, for a single plant obviously cannot yield enough grain to mill and bake. This fact no doubt explains the differences of opinion among plant breeders on the inheritance of strength, for it is not every one who can acquire the power of judging wheat accurately by his senses. Such a faculty is a personal gift, and is at best apt to fail at times.

The search for a rapid and accurate method of measuring strength has for many years attracted the attention of investigators. As might be expected

IV] QUALITY OF WHEAT: MILLER'S VIEW 63

most of the investigations have centred round the gluten, for as mentioned above the gluten of a strong wheat is much more tough and elastic than that of a weak wheat. Gluten is a characteristic constituent of all wheats, and it is the presence of gluten which gives to wheat flour the power of making bread. The other cereals, barley, oats, maize and rice are very similar to wheat in their general chemical composition, but they do not contain gluten. Consequently they cannot make bread.

In making bread flour is mixed with water and yeast. The yeast feeds on the small quantity of sugar contained in the flour, fermenting it and forming from it alcohol and carbon dioxide gas. The gluten being coherent and tough is blown into numberless small bubbles by the gas, which is thus retained inside the bread. On baking, the high temperature of the oven fixes these bubbles by drying and hardening their walls, and the bread is thus endowed with its characteristic porous structure. If a cereal meal devoid of gluten is mixed with water and yeast, fermentation will take place with formation of gas, but the gas will escape at once, and the product will be solid and not porous. Evidently from the baking point of view gluten is of the greatest importance. One of the most obvious methods that have been suggested for estimating the strength of wheat depends on the estimation of the percentage of gluten contained in the flour. The method has not turned out very successfully, for strength seems to depend rather on the quality than on the quantity of gluten in the wheat. Much attention has been given to the study of the causes of the varying quality of the gluten of different wheats. Gluten for instance has been shown to be a mixture of two substances, gliadin and glutenin, and the suggestion has been made that its varying properties are dependent on the varying proportions of these two substances present in different samples. This suggestion however failed to solve the problem.

After seven years of investigation the author has worked out the following theory of the strength of wheat flours, which has finally enabled him to devise a method which promises to be both accurate and rapid, and to require so little flour that it can readily be used by the wheat breeder to determine the strength of the grain in a single ear. It has already been mentioned that a strong wheat is one that will make a large loaf of good shape and texture. The strength of a wheat may therefore be defined as the power of making a large loaf of good shape and texture. Evidently strength is a complex of at least two factors, size and shape, which are likely to be quite independent of each other. Not infrequently, for instance, wheats are met with which make large loaves of bad shape, or on the other hand, small

IV] QUALITY OF WHEAT: MILLER'S VIEW

loaves of good shape. Probably therefore the size of the loaf depends on one factor, the shape on another; and the failure of the many attempts to devise a method of estimating strength have been caused by the impossibility of measuring the product of two independent factors by one measurement.

It seemed a feasible idea that the size of the loaf might depend on the volume of gas formed when yeast was mixed with different flours. On mixing different flours with water and yeast it was found that for the first two or three hours they all gave off gas at about the same rate. The reason of this is that all flours contain about the same amount of sugar, approximately one per cent., so that at the beginning of the bread fermentation all flours provide the yeast with about the same amount of sugar for food. But this small amount of sugar is soon exhausted, and for its subsequent growth the yeast is dependent on the transformation of some of the starch of the flour into sugar. Wheat like many other seeds contains a ferment or enzyme called diastase, which has the power of changing starch into sugar, and the activity of this ferment varies greatly in different wheats. The more active the ferment in a flour the more rapid the formation of sugar. Consequently the more rapidly the yeast will grow, and the greater will be the volume of gas produced in the later stages of fermentation in the dough. As a rule it is not practicable to get the dough moulded into loaves and put into the oven before it has been fermenting for about six or eight hours. If the flour possesses an active ferment it will still be rapidly forming gas at the end of this time, and the loaf will go into the oven distended with gas under pressure from the elasticity of the gluten which forms the walls of the bubbles. The heat of the oven will cause each gas bubble to expand, and a large loaf will be the result. If the ferment of the flour is of low activity it will not be able to keep the yeast supplied with all the sugar it needs, the volume of gas formed in the later stages of the fermentation of the dough will be small, the dough will go into the oven without any pressure of gas inside it, little expansion will take place as the temperature rises, and a small loaf will be produced.

From these facts it is quite easy to devise a method of estimating how large a loaf any given flour will produce. The following method is that used by the author. A small quantity of the flour, usually 20 grams, is weighed out and put into a wide mouthed bottle. A flask of water is warmed to 40° C., of this 100 c.c. is measured out, and into it $2\frac{1}{2}$ grams of compressed yeast is intimately mixed, 20 c.c. of the mixture being added to the 20 grams of flour in the bottle. The flour and yeast-water are then mixed

IV] QUALITY OF WHEAT: MILLER'S VIEW

into a cream by stirring with a glass rod. The bottle is then placed in a vessel of water which is kept by a small flame at 35° C. The bottle is connected to an apparatus for measuring gas, and the volume of gas given off every hour is recorded. As already mentioned all flours give off about the same volume of gas during the first three hours. After this length of time the volume of gas given off per hour varies greatly with different flours. Thus a flour which will bake a large loaf gives off under the conditions above described about 20 c.c. of gas during the sixth hour of fermentation, whilst a flour which bakes a small tight loaf gives off during the sixth hour of fermentation only about 5 c.c. of gas.

Having devised a feasible method of estimating how large a loaf any given flour will make, the problem of the shape and texture still remains. Previous investigators had exhausted almost every possible chemical property of gluten in their search for a method of estimating strength. The author therefore determined to study its physical properties. Now gluten is what is known as a colloid substance, like albumen the chief constituent of white of egg, casein the substance which separates when milk is curdled, or clay which is a well known constituent of heavy soils. Such colloid substances can scarcely be said to possess definite physical properties of their own, for their properties vary so largely with their surroundings. The white of a fresh egg is a thick glairy liquid. On heating it becomes a white opaque solid, and the addition of certain acids produces a similar change in its properties. Casein exists in fresh milk in solution. The addition of a few drops of acid causes it to separate as finely divided curd. If, however, the milk is warmed before the acid is added the casein separates as a sticky coherent mass. Every farmer knows that lime improves the texture of soils containing much clay, because the lime causes the clay to lose its sticky cohesive nature.

Such instances show that the properties of colloid substances are profoundly modified by the presence of chemical substances. Wheat, like almost all plant substances, is slightly acid, and the degree of acidity varies in different samples. Accordingly the effect of acids on the physical properties of gluten was investigated, and it was found that by placing bits of gluten in pure water and in acid of varying concentration it could be made to assume any consistency from a state of division so fine that the separate particles could not be seen, except by noticing that their presence made the water milky, to a tough coherent mass almost like indiarubber (Fig. 11). It was found, however, that the concentration of acid in the wheat grain was never great enough to make the gluten really coherent.

Fig. 11.

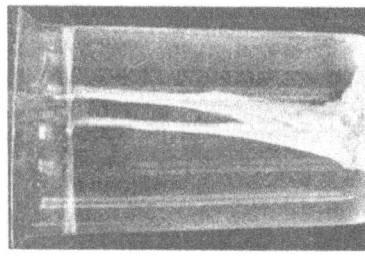

Gluten in pure water; soft, but tough and elastic

Gluten in very weak hydrochloric acid (3 parts in 100,000 of water); it floats about in powder, having entirely lost cohesion, and makes the water milky

Gluten in hydrochloric acid (3 parts in 1000 of water); very hard and tough

But wheat contains also varying proportions of such salts as chlorides, sulphates and phosphates, which are soluble in water, and the action of such salts on gluten was next tried. It was at once found that these salts in the same concentration as they exist in the wheat grain were capable of making gluten coherent, but that the kind of coherence produced was peculiar to each salt. Phosphates produce a tough and elastic gluten such as is found in the strongest wheats. Chlorides and sulphates on the other hand make gluten hard and brittle, like the gluten of a very weak wheat (Fig. 12).

The next step was to make chemical analyses to find out the amount of soluble salts in different wheats. Strong wheats of the Fife class were found to contain not less than 1 part of soluble phosphate in 1000 parts of wheat, whilst Rivet wheat, the weakest wheat that comes on the market, contained only half that amount. Rivet, however, was found to be comparatively rich in soluble chlorides and sulphates, which are present in very small amounts in strong wheats of the Fife class. Ordinary English wheats resemble Rivet, but they contain rather more phosphate and rather less chlorides and sulphates. After making a great many analyses it was found that the amount of soluble phosphate in a wheat was a very good index of the shape and texture of the loaf which it would make. The toughness and elasticity

Fig. 12.

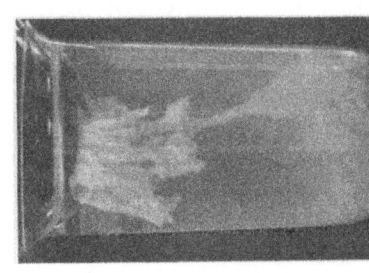
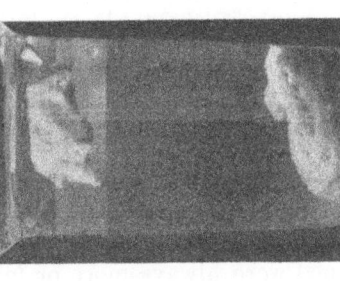
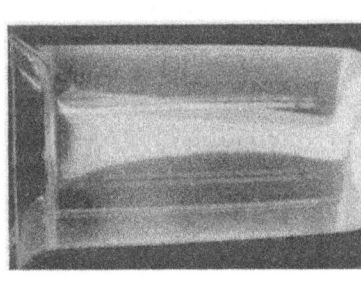

Gluten in water containing both acid and phosphate; very tough and elastic

Gluten in water containing both acid and sulphates. It shows varying degrees of coherence, but is brittle or "short"

of the gluten no doubt depend on the concentration of the soluble phosphate in the wheat grain, the more the soluble phosphate the tougher and more elastic the gluten, and a tough and elastic gluten holds the loaf in shape as it expands in the oven, and prevents the small bubbles of gas running together into large holes and spoiling the texture.

These facts suggest at once a method for estimating the shape and texture of the loaf which can be made from any given sample of wheat. An analysis showing the amount of soluble phosphate in the sample should give the desired information. But unfortunately such an analysis is not an easy one to make, and requires a considerable quantity of flour. In making these analyses it was noticed that when the flours were shaken with water to dissolve the phosphate, and the insoluble substance removed by filtering, the solutions obtained were always more or less turbid, and the degree of turbidity was found to be related to the amount of phosphate present and to the shape of loaf produced. On further investigation it was found that the turbidity was due to the fact that the concentration of acid and salts which make gluten coherent, also dissolve some of it, and gluten like other colloids gives a turbid solution. It was also found that the amount of gluten dissolved, and consequently the degree of turbidity, is related to the shape of the loaf which the flour will produce.

IV] QUALITY OF WHEAT: MILLER'S VIEW 73

Now it is quite easy to measure the degree of turbidity of a solution by pouring the solution into a glass vessel below which a small electric lamp is placed, and noting the depth of the liquid through which the filament of the lamp can just be seen. The turbidities were, however, so slight that it was found necessary to increase them by adding a little iodine solution which gives a brown milkiness with solutions of gluten, the degree of milkiness depending on the amount of gluten in the solution. In this way a method was devised which is rapid, easy, and can be carried out with so little wheat that the produce of one ear is amply sufficient. It can therefore be used by the plant breeder for picking out from the progeny of his crosses those individual plants which are likely to give shapely loaves. The method is as follows: An ear of wheat is rubbed out and ground to powder in a small mill. One gram of this powder, or of flour if that is to be tested, is weighed out and put into a small bottle. To it is added 20 c.c. of water. The bottle is then shaken for one hour. At the end of this time the contents are poured onto a filter. To 15 c.c. of the solution $1\frac{1}{2}$ c.c. of a weak solution of iodine is added, and after standing for half an hour the turbidity test is applied. Working in this way it is possible to see through only 10 c.m. of the solution thus obtained from such a wheat as Red Fife, as compared with 25 c.m. in the case of Rivet.

Other wheats yield solutions of intermediate opacity. This method is now being tested in connection with the Cambridge wheat breeding experiments.

CHAPTER V

THE MILLING OF WHEAT

In order that wheat may be made into bread it is necessary that it should be reduced to powder. In prehistoric times this was effected by grinding the grain between stones. Two stones were commonly used, the lower one being more or less hollowed on its upper surface so as to hold the grain while it was rubbed by the upper one. As man became more expert in providing for his wants, the lower stone was artificially hollowed, and the upper one shaped to fit it, until in process of time the two stones assumed the form of a primitive mortar and pestle.

The next step in the evolution of the mill was to make a hole or groove in the side of the lower stone through which the powdered wheat could pass as it was ground. This device avoided the trouble of emptying the primitive mill, and materially saved the labour of the grinder. Such mills are still in use in the less civilised countries in the East, and are of course worked by hand as in primitive times.

They gradually developed as civilization progressed

THE MILLING OF WHEAT

into the stone mills which ground all the breadstuffs of the civilised world until about 40 years ago. The old fashioned stone mill was, and indeed still is, a weapon of the greatest precision. It consists of a pair of stones about four feet in diameter, the lower of which is fixed whilst the upper is made to revolve by mechanical power at a high speed. Each stone is made of a large number of pieces of a special kind of hard stone obtained from France. These pieces are cemented together, and the surfaces which come into contact are patiently chipped until they fit one another to a nicety all over. The surface of the lower stone is then grooved so as to lead the flour to escape from between the stones at definite places where it is received for further treatment. The grain to be ground is fed between the stones through a hole at the centre of the upper stone. It has been stated above that the surfaces of the two stones are in contact. As a matter of fact this is not strictly true. The upper stone is suspended so that the surfaces are separated by a small fraction of an inch, and it will be realised at once that this suspension is a matter of the greatest delicacy. To balance a stone weighing over half a ton so that, when revolving at a high rate of speed, it may be separated from its partner at no point over its entire surface of about 12 square feet by more than the thickness of the skin of a grain of wheat, and yet may nowhere come into actual contact, is an achievement of no mean order. Stone mills of this kind were usually driven by water power, or in flat neighbourhoods by wind power, though in some cases steam was used.

It was the common practice to subject the ground wheat from the stones to a process of sifting so as to remove the particles of husk from the flour. The sifting was effected by shaking the ground wheat in a series of sieves of finely woven silk, known as bolting cloth. In this way it was possible to obtain a flour which would make a white bread. The particles of husk removed by the sifting were sold to farmers for food for their animals, under the name of bran, sharps, pollards, or middlings, local names for products of varying degrees of fineness, which may be classed together under the general term wheat offals. The ideal of the miller was to set his stones so that they would grind the flour to a fine powder without breaking up the husk more than was absolutely necessary. When working satisfactorily a pair of stones were supposed to strip off the husk from the kernel. The kernel should then be finely pulverised. The husk should be flattened out between the stones, which should rub off from the inside as completely as possible all adhering particles of kernel. If this ideal were attained, the mill would yield a large proportion of fairly white flour, and a small proportion of husk or offals.

THE MILLING OF WHEAT

As long as home grown wheats were used this ideal could be more or less attained because the husk of these wheats is tough and the kernel soft. Comparatively little grinding suffices to reduce the kernel to the requisite degree of fineness, and this the tough husk will stand without being itself unduly pulverised. Consequently the husk remains in fairly large pieces, and can be separated by sifting, with the result that a white flour can be produced. But home grown wheat ceased to provide for the wants of the nation more than half a century ago. Already in 1870 half the wheat ground into flour in the United Kingdom was imported from abroad, and this proportion has steadily increased, until at the present time only about one-fifth of the wheat required is grown at home. Many of the wheats which are imported are harder in the kernel, and thinner and more brittle in the husk, than the home grown varieties. Consequently they require more grinding to reduce the kernel to the requisite degree of fineness, and their thin brittle husk is not able to resist such treatment. It is itself ground to powder along with the kernel, and cannot be completely separated from the flour by sifting. Such wheats therefore, when ground between stones, yield flour which contains much finely divided husk, and this lowers its digestibility and gives it a dark colour.

In the decades before 1870 when the imports of

foreign wheats first reached serious proportions, and all milling was done by stones, dark coloured flours were common, and people would no doubt have accepted them without protest, if no other flours had been available. But as it happened millers in Hungary, where hard kernelled, thin skinned wheats had long been commonly grown, devised the roller milling process, which produces fine white flour from such wheats, no matter how hard their kernels or how thin their skins. The idea of grinding wheat between rollers was at once taken up in America and found to give excellent results with the hard thin skinned wheats of the north-west. The fine white flours thus produced were sent to England, and at once ousted from the home markets the dark coloured flours produced from imported wheats in the English stone mills. The demand for the white well-risen bread produced from these roller milled imported flours showed at once that the public preferred such bread to the darker coloured heavier bread yielded by stone-ground flours, especially those made from the thin skinned foreign wheats.

This state of things was serious both for the millers and the farmers. The importation of flour instead of wheat must obviously ruin the milling industry, and since wheat offals form no inconsiderable item in the list of feeding stuffs available for stock keepers, a decline of the milling industry restricts the supply of food for his stock, and thus indirectly affects the farmer. At the same time the preference shown by the public for bread made from fine white imported flour, to some extent depreciated the value of home grown wheat.

It was by economic conditions of this kind that the millers were compelled in the early seventies to alter their methods. The large firms subscribed more capital and installed roller plant in their mills. These at once proved a success and the other firms have followed suit. At the present time considerably more than 90 per cent. of the flour used in this country is the product of roller mills. The keen competition which has arisen in the milling industry during the last 35 years has produced great improvements in roller plant, and the methods of separation now in use yield flours which in the opinion of the miller, and apparently too in the opinion of the general public, are far in advance of the flours which were produced in the days of stone milling.

Perhaps the first impression which a visitor to a modern roller mill would receive is the great extent to which mechanical contrivances have replaced hand labour. Once the wheat has been delivered at the mill it is not moved again by hand until it goes away as flour and offals. It is carried along by rapidly moving belts, elevated by endless chains carrying buckets, allowed to fall again by gravity, or perhaps in other cases transported by air currents. Another very striking development is the great care expended in cleaning the grain before it is ground. This cleaning is the first process to which the wheat is subjected. It is especially necessary in the case of some of the foreign wheats which arrive in this country in a very dirty condition. The impurities consist of earth, weed seeds, bits of husk and straw; iron nails, and other equally unlikely objects are by no means uncommon. Some of these are removed by screens, but besides screening the wheat is actually subjected to the process of washing with water. For this purpose it is elevated to an upper floor of the mill, and allowed to fall downwards through a tall vessel through which a stream of water is made to flow. As it passes through the water it is scrubbed by a series of mechanically driven brushes to remove the earthy matter which adheres to the grain. This is carried away by the stream of water.

After cleaning the grain next undergoes the process of conditioning. The object of this process is so to adjust the moisture of the grain that the husk may attain its maximum toughness compatible with a reasonable degree of brittleness of kernel, the idea being to powder the kernel with the minimum of grinding and without unduly powdering the husk. By attention to this process separation of flour and

v] THE MILLING OF WHEAT 81

husk is made easier and more complete. The essential points in the process are to moisten the grain, either in the course of cleaning as above described, or if washing is not necessary, by direct addition of water. The moisture is given some time to be absorbed into the grain, which is then dried until the moisture content falls to what experience shows to be the most successful figure for the wheat in question.

Fig. 13. First break rolls seen from one end. The ribs can just be seen where the two rolls touch

Cleaning and conditioning having been attended to, the grain is now conveyed to the mill proper. This of course is done by a mechanical arrangement which feeds the grain at any desired rate into the hopper which supplies the first pair of rolls. These rolls consist of a pair of steel cylinders usually

10 inches in diameter and varying in length from 20 inches to 5 feet according to the capacity of the mill. The surfaces of the cylinders are fluted or ribbed, the distance from rib to rib being about one-tenth of an inch. The rollers are mounted so

Fig. 14. Break rolls. The large and small cog-wheels are the simplest device used to give the two rolls different speeds. The larger cog-wheel is driven by power and drives the smaller, of course at a much higher rate of revolution

that the distance between their surfaces can be adjusted. They are set so that they will break grains passing between them to from one-half to one-quarter their original size. They are made to

THE MILLING OF WHEAT

revolve so that the parts of the surfaces between which the grains are nipped are travelling in the same direction. One roll revolves usually at about 350 revolutions per minute, the other at rather less than half that rate (Fig. 14). It is obvious from the above description that a grain of wheat falling from the hopper on to the surface of the moving rollers will be crushed or nipped between them, and that since the rollers are moving at different rates, it will at the same time be more or less torn apart. By altering the distance between the rollers and their respective speeds of revolution the relative amounts of nipping and tearing can be adjusted to suit varying conditions.

The passage of the grain through such a pair of rollers is known technically as a break. Its object is to break or tear open the grain with the least possible amount of friction between the grain and the grinding surfaces. Since the rollers are cylindrical it is obvious that the grain will only be nipped at one point of their surfaces, and even here the friction is reduced as much as possible by making both the grinding surfaces move in the same direction. As already explained it can be diminished, if the condition of the wheat allows, by diminishing the difference in speed between the two rolls. The result of the first break is to tear open the grains. At the same time a small amount of the kernel will be finely powdered. The rest of the kernel and husk will still remain in comparatively large pieces. The tearing open of the grain sets free the dirt which was lodged in the crack or furrow which extends from end to end of the grain. This dirt cannot be removed by any method of cleaning. It only escapes when the grain is torn open in the break. It is generally finely divided dirt and cannot be separated from the flour formed in this process. Consequently the first break flour is often more or less dirty, and the miller tries to adjust his first break rolls so that they will form as little flour as possible. The first break rolls not only powder a little of the kernel, but they also reduce to a more or less fine state of division a little of the husk.

The result of the passage of the grain through the first break rolls is to produce from it a mixture of a large quantity of comparatively coarse particles of kernel to many of which husk is still adherent, a small quantity of finely divided flour which is more or less discoloured with dirt, and a small quantity of finely divided husk. This mixture, which is technically known as stock, is at once subjected to what is called separation, with the object of separating the flour from the other constituents before it undergoes any further grinding. It is one of the guiding principles of modern milling that the flour produced at each operation should be separated at once so as to

THE MILLING OF WHEAT

reduce to a minimum the grinding which it has to undergo. Separation is brought about by the combination of two methods. The stock is shaken in contact with a screen made of bolting silk so finely woven that it contains from 50 to 150 meshes to the inch, according to the fineness of the flour which it is desired to separate. The shaking is effected in several different ways. Sometimes the silk is stretched on a frame so as to make a kind of flat sieve. This is shaken mechanically whilst the stock is allowed to trickle over its surface, so that the finely divided particles of flour may fall through the meshes and be collected separately from the larger particles which remain on the top. These larger particles are partly heavy bits of broken kernel and partly light bits of torn husk. In order to separate them advantage is taken of the fact that a current of wind can be so adjusted that it will blow away the light and fluffy husk particles without disturbing the heavy bits of kernel. By means of a mechanically driven fan a current of air is blown over the surface of the sieve, in the direction opposite to that in which the stock is travelling. As the stock rolls over and over in its passage from the upper to the lower end of the inclined sieve the fluffy particles of husk are picked up by the air current and carried back to the top of the sieve where they fall, as the current slackens, into a receptacle placed to receive them. Thus by the combination of sifting and air carriage the stock is separated into a small quantity of finished flour, a small quantity of finished husk or offal, and a large quantity of large particles of kernel with husk still adhering to some of them. These large particles, which are called semolina, of course require further grinding. Different methods of sifting are often used in place of the one above described, especially for completing the purification of the flour. Sometimes the silk is stretched round a more or less circular frame so as to form a long cylinder covered with silk. The stock is delivered into the higher end of this cylinder which is made to revolve. This causes the stock to work its way through the cylinder, and during its progress the finely ground flour finds its way through the meshes, and is separated as before from the coarser particles. Such a revolving sieve is known as a reel. In a somewhat similar arrangement known as a centrifugal a series of beaters is made to revolve rapidly inside a stationary cylindrical sieve. The stock is admitted at one end and is thrown by the revolving beaters against the silk cover. The finer particles are driven through the meshes of the silk, the coarser particles find their way out of the cylinder at the other end. Sometimes for separating very coarse particles wire sieves of 30 meshes, or thereabouts, to the inch are used. Whatever the method the object is to separate at once the finished

THE MILLING OF WHEAT

flour and offal from the large particles of kernel which require further grinding.

These large particles, semolina, are next passed between one or more pairs of smooth rolls known as reduction rolls (Fig. 15). These are set rather nearer together than the break rolls, and the difference in speed between each roll and its partner is quite

Fig. 15. A pair of reduction rolls. They are smooth, and the cog-wheels being nearly of the same size the speed of the two rolls is nearly equal

small. The object of reduction is to reduce the size of the large particles of semolina and to produce thereby finely divided flour. The stock from the first pair or pairs of reduction rolls contains much finely ground flour mixed with coarser particles of kernel with or without adherent husk. It is at once submitted to the separation and purification processes as above described. This yields a large quantity of finished flour which is very white and free from husk. It represents commercially the highest grade of flour separated in the mill and is described technically as patents. A small amount of finished offal is also separated at this stage.

The coarse particles of kernel with adherent husk from which the flour and offal have been separated are now passed through a second pair of break rolls more finely fluted than before, known as the second break. These are set closer together than the first break rolls. Their object is to rub off more kernel from the husk. The stock from them is again separated, the flour and finished offal being removed as before. The coarser particles are again reduced by smooth reduction rolls, and a second large quantity of flour separated. This is commercially high grade flour and is usually mixed with the patents already separated. The coarse particles left after this separation are usually subjected to a third and a fourth break, each of which is succeeded by one or two reductions. Separation of the stock and purification of the flour take place after each rolling, so that as soon as any flour or husk is finely ground it may be at once separated without further grinding. The last pair of fluted rolls, the fourth break, are set so closely together that they practically touch both sides of the pieces of husk which pass through them. They are intended to scrape the last particles of kernel from the husk. This is very severe treatment, and usually results in the production of much finely powdered husk which goes through the sifting silk and cannot be separated from the flour. The flour from the fourth break is therefore usually discoloured by the presence of much finely divided husk. For this reason it ranks as of low commercial grade. The later reductions too yield flours containing more or less husk, which darkens their colour. They are usually mixed together and sold as seconds.

The fate of the germ in the process of roller milling is a point of considerable interest, both on account of the ingenious way in which it is removed, and because of the mysterious nutritive properties which it is commonly assumed to possess. The germ of a grain of wheat forms only about $1\frac{1}{2}$ per cent. by weight of the grain. It differs in composition from the rest of the grain, being far richer in protein, fat, and phosphorus. Its special feeding value can, however, scarcely be explained in terms of these ingredients, for its total amount is so small that its presence or absence in the flour can make only a very slight difference in the percentages of these substances. But this point will be discussed fully in a subsequent chapter. Here it is the presence of the fat which is chiefly of interest. According to the millers the fat of the germ is prone to become rancid, and to impart to the flour, on keeping, a peculiar taste and odour which affects its commercial value. They have therefore devised with great ingenuity a simple method of removing it. This method depends on the fact that the presence in the germ of so much fat prevents it from being ground to powder in its passage between the rolls. Instead of being ground it is pressed out into little flat discs which are far too large to pass with the flour through the sifting silks or wires, and far too heavy to be blown away by the air currents which remove the offals. The amount which is thus separated is usually about 1 per cent. of the grain so that one third of the total quantity of germ present in the grain is not removed as such. Considerable difficulties arise in attempting to trace this fraction, and at present it is impossible to state with certainty what becomes of it. The germ which is separated is sold by the ordinary miller to certain firms which manufacture what are known as germ flours. It is subjected to a process of cooking which is said to prevent it from going rancid, after which it is ground with wheat, the product being patent germ flour.

CHAPTER VI

BAKING

In discussing the method of transforming flour into bread it will be convenient to begin by describing in detail one general method. The modifications used for obtaining bread of different kinds, and for dealing with flours of different qualities will be shortly discussed later when they can be more readily understood.

Bread may be defined as the product of cooking or baking a mixture of flour, water, and salt, which is made porous by the addition of yeast. It is understood to contain no other substances than these—flour, salt, water and yeast.

In the ordinary process the first step is to weigh out the flour which it is proposed to bake. This is then transferred to a vessel which in a commercial bakery is usually a large wooden trough, in a private house an earthenware bowl. The necessary amount of yeast is next weighed out and mixed with water. Nowadays compressed or German yeast is almost always used at the rate of 1 to 2 lbs. per sack or 280 lbs. of flour. For smaller quantities of flour relatively more yeast is needed, for instance 2 ozs. per stone. Formerly brewers' yeast or barm was used, but its use has practically ceased because it

92 THE STORY OF A LOAF OF BREAD [CH.

is difficult to obtain of standard strength. Some people who profess to be connoisseurs of bread still prefer it because as they say it gives a better flavour to the bread. The water with which the yeast is mixed is warmed so as to make the yeast more active. The

Fig. 16. Apparatus arranged for a baking test. Four loaves which have just been scaled and moulded are seen in an incubator where they are left to rise or prove before being transferred to the oven

flour is then heaped up at one end of the vessel in which the mixing is to take place, and salt at the rate of 2 to 5 lbs. per sack is thoroughly stirred into it. A hollow is then made in the heap of flour into which the mixture of yeast and water is poured.

VI] BAKING 93

More warm water is added so that enough water in all may be present to convert all, or nearly all, the flour into dough of the required consistency. When dealing with a flour with which he is familiar the baker knows by experience how much water he

Fig. 17. The loaves shown in the last figure have just been baked and are ready to be taken out of the oven, the door of which is open. Note the different shapes. That on the right hand is obviously shown by the test to be made from a strong flour, the other from a very weak flour

requires per sack. In the case of an unaccustomed brand of flour he determines the amount by a preliminary trial with a small quantity (Figs. 16 and 17). Flour from the heap is then stirred into the water until the whole of the flour is converted into a stiff paste or dough as it is called. By this method a little dry flour will always separate the dough from the sides of the vessel and this will prevent the dough from sticking to the vessel and the hands. The dough is then thoroughly worked or kneaded so as to ensure the intimate mixture of the ingredients. The vessel is then covered to keep the dough warm. In private houses this is ensured by placing the vessel near the fire. In bakeries the room in which the mixing is conducted is usually kept at a suitable temperature. The yeast cells which are thoroughly incorporated in the dough, find themselves in possession of all they require to enable them to grow. The presence of water keeps them moist, and dissolves from the flour for their use sugar and salts: the dough is kept warm as above explained. Under these conditions active fermentation takes place with the formation of alcohol and carbon dioxide gas. The alcohol is of no particular consequence in bread making, the small amount formed is probably expelled from the bread during its stay in the oven. The carbon dioxide, however, plays a most important part. Being a gas it occupies a large volume, and its formation throughout the mass of the dough causes the dough to increase greatly in volume. The dough is said by the housewife to rise, by the professional baker to prove.

The process of kneading causes the particles of gluten to absorb water and to adhere to one another, so that the dough may be regarded as being composed of innumerable bubbles each surrounded by a thin film of gluten, in or between which lie the starch grains and other constituents of the flour. Each yeast cell as above explained forms a centre for the formation of carbon dioxide gas, which cannot escape at once into the air, and must therefore form a little bubble of gas inside the particular film of gluten which happens to surround it. The expansion of the dough is due to the formation inside it of thousands of these small bubbles. It is to the formation of these bubbles too that the porous honeycombed structure of wheaten bread is due. Also since the formation of the bubbles is due to the retention of the carbon dioxide by the gluten films, such a porous structure is impossible in bread made from the flour of grains which do not contain gluten.

The rising of the dough is usually allowed to proceed for several hours. The baker finds by experience how long a fermentation is required to give the best results with the flours he commonly uses. When the proper time has elapsed, the dough is removed from the trough or pan in which it was mixed to a board or table, previously dusted with dry flour to prevent the dough adhering to the board or to the hands. It is then divided into portions of the proper weight to make loaves of the desired size. This process is known technically as scaling. Usually 2 lbs. 3 ozs. of dough is allowed for baking a 2 lb. loaf. Each piece of dough is now moulded into the proper shape if it is desired to bake what is known as a cottage loaf, or placed in a baking tin if the baker is satisfied with a tinned loaf. In either case the dough is once more kept for some time at a sufficiently warm temperature for the yeast to grow so that the dough may once more be filled with bubbles of carbon dioxide gas. As soon as this second rising or proving has proceeded far enough the loaves are transferred to the oven. Here the intense heat causes the bubbles of gas inside the dough to expand so that a sudden increase in the size of the loaf takes place. At the same time the outside of the loaf is hardened and converted into crust.

After remaining in the oven for the requisite time the bread is withdrawn and allowed to cool as quickly as possible, after which it is ready for use or sale.

The method of baking which has been described above is known as the off-hand or straight dough method. It possesses the merit of rapidity and simplicity, but it is said by experts that it does not yield the best quality of bread from certain flours. Perhaps the commonest variation is that known as the sponge and dough method, which is carried out as follows.

BAKING

As before, the requisite amount of flour is weighed out into the mixing trough, and a depression made in it for the reception of the water and yeast. These are mixed together in the proper proportions, enough being taken to make a thick cream with about one quarter of the flour. This mixture is now poured into the depression in the flour, and enough of the surrounding flour stirred into it to make a thick cream or sponge as it is called. At the same time a small quantity of salt is added to the mixture. The sponge is allowed to ferment for some hours, being kept warm as in the former method. As soon as the time allowed for the fermentation of the sponge has elapsed, more water is added, so that the whole or nearly the whole of the flour can be worked up into dough. This dough is immediately scaled and moulded into loaves, which after being allowed to prove or rise for some time are baked as before. This method is used for flours which do not yield good bread when they are submitted to long fermentation. In such cases the mellow flours, which will only stand a very short fermentation, are first weighed out into the mixing trough, and a depression made in the mass of flour into which a quantity of strong flour which can be fermented safely for a long time is added. It is this last addition which is mixed up into the sponge to undergo the long preliminary fermentation. The rest of the flour is mixed in after this first fermentation is over, so that it is only subjected to the comparatively slight fermentation which goes on in the final process of proving.

Many other modifications are commonly practised locally, their object being for the most part to yield bread which suits the local taste. It will suffice to mention one which has a special interest. In this method the essentially interesting point is the preparation of what is known as a ferment. For this purpose a quantity of potatoes is taken, about a stone to the sack of flour. After peeling and cleaning they are boiled and mashed up with water into a cream. To this a small quantity of yeast is added and the mixture kept warm until fermentation ceases, as shown by the cessation of the production of gas. During this fermentation the yeast increases enormously, so that a very small quantity of yeast suffices to make enough ferment for a sack of flour. The flour is now measured out into the trough, and the ferment and some additional water and salt added so that the whole can be worked up into dough. Scaling, moulding, and baking are then conducted as before. This method was in general use years ago when yeast was dear. It has fallen somewhat into disuse in these days of cheap compressed yeast, in fact the use of potatoes nowadays would make the process expensive.

In private houses and in the smaller local bakeries the whole of the processes described above are carried out by hand. During the last few decades many very large companies have been formed to take up the production of bread on the large scale. This has caused almost a revolution of the methods of manipulating flour and dough, and in many cases nowadays almost every process in the bakery is carried out by machinery. In many of the larger bakeries doughing and kneading are carried out by machines, and this applies also to the processes of scaling and moulding. A similar change has taken place too in the construction of ovens. Years ago an oven consisted of a cavity in a large block of masonry. Wood was burned in the cavity until the walls attained a sufficiently high temperature. The remains of the fuel were then raked out and the bread put in and baked by radiation from the hot walls.

Nowadays it is not customary to burn fuel in the oven itself, nor is the fuel always wood or even coal. The fuel is burned in a furnace underneath the oven, and coal or gas is generally used. Sometimes however the source of heat is electricity. In all cases it is still recognised that the heat should be radiated from massive solid walls maintained at a high temperature. In the latest type of oven the heat is conducted through the walls by closed iron tubes containing water, which of course at the high temperatures employed becomes superheated steam. It is recognised that the ovens commonly provided in modern private houses, whether heated by the fire of the kitchen range, or by gas, are not capable of baking bread of the best quality, because their walls do not radiate heat to the same degree as the massive walls of a proper bake oven.

It is commonly agreed that bread, in the usual acceptation of the term, should contain nothing but flour, yeast, salt, and water; or if other things are present they should consist only of the products formed by the interaction of these four substances in the process of baking. Millers and bakers have, however, found by experience that the addition of certain substances to the flour or to the dough may sometimes enable them substantially to improve the market value of the bread produced by certain flours. The possible good or bad effect of such additions on the public health will be discussed in a later chapter. It may be of interest here to mention some of the substances which are commonly used as flour or bread improvers by millers and bakers, and to discuss the methods by which they effect their so called improvements.

In a former chapter we have discussed the quality of wheat from the miller's point of view, and during the discussion certain views were enunciated on the subject of strength. It was pointed out that a strong flour was one which would make a large well-shaped loaf, and that the size of the loaf was dependent on the flour being able to provide sugar for the yeast to feed upon right up to the moment when the loaf goes into the oven. This can only occur when the flour contains an active ferment which keeps changing the starch into sugar. That this view is generally accepted in practice is shown by the fact that, when using flours deficient in such ferment, bakers commonly add to the flour, yeast, salt, and water, a quantity of malt extract, the characteristic constituent of which is the sugar producing ferment of the malt. This use of malt extract is now extending to the millers, several of whom have installed in their mills plant for spraying into their flour a strong solution of malt extract. It seems to be agreed by millers and bakers generally that such an addition to a flour which makes small loaves distinctly increases the size of the loaf. There can be no doubt that this effect is produced by the ferment of the malt extract keeping up the supply of sugar, and thus enabling the yeast to maintain the pressure of gas in the dough right up to the moment when it goes into the oven.

The view that the shape of the loaf is due to the effect of salts, and particularly of phosphates, on the coherence of the gluten has also been put to practical use by the millers and the bakers. Preparations of phosphates under various fancy names are now on the market, and are bought by bakers for adding to the flour to strengthen the gluten and produce more shapely loaves. A few millers too are beginning to spray solutions of phosphates into their flours with the same object in view, and such additions are said to make material improvements in the shape of the loaf produced by certain weak flours.

These two substances, malt extract and phosphates, are added to the flour with the definite object of improving the strength and thus making larger and more shapely loaves. But there is a second class of substances which are commonly added to flours, not in the mill but in the process of bread making, with the object of replacing yeast. Yeast is used in baking in order that it may form gas inside the dough and thus produce a light spongy loaf. Exactly the same gas can be readily and cheaply produced by the interaction of a carbonate with an acid. These substances will not react to produce acid as long as they remain dry, but as soon as they are brought into close contact with each other by the presence of water, reaction begins and carbon dioxide gas is formed. These facts are taken advantage of in the manufacture of baking powders and self-rising flours. Baking powders commonly consist of ordinary bicarbonate of soda mixed with an acid or an acid salt, such as tartaric acid, cream of tartar, acid phosphate of lime, or acid phosphate of potash. One of these latter acid substances is mixed in proper proportions with the bicarbonate of soda, and the mixture ground up with powdered starch which serves to dilute the chemicals and to keep them dry. A small quantity of the baking powder is mixed with the flour before the water is added to make the dough. The presence of the water causes the acid and the carbonate to give off gas which, as in the case of the gas formed by the growth of yeast, fills the dough with bubbles which expand in the oven and produce light spongy bread. When using baking powders in place of yeast it must not be forgotten that gas formation in most cases begins immediately the water is added, and lasts for a very short time. Consequently the dough must be moulded and baked at once or the gas will escape. This is not the case, however, with those powders which are made with cream of tartar, for this substance does not react with the carbonate to any great extent until the dough gets warm in the oven. For some purposes it is customary to use carbonate of ammonia, technically known as volatile, in place of baking powder. This substance is used alone without any addition of acid, because it decomposes when heated and forms gas inside the dough. Sometimes too one or other of the baking powders above described are added to the flour by the miller, the product being sold as self-rising flour. Such flour will of course lose its property of self-rising if allowed to get damp. Occasionally objectionable substances are used in making baking powders of self-rising flours. Some baking powders for instance contain alum which is not a desirable addition to any article of human food. Baking powders and self-rising flours are far more frequently used by house-wives for making pastry or for other kinds of domestic cookery than for breadmaking.

Bread is made on the large scale without the intervention of yeast by the aeration process, which is carried out as follows. A small quantity of malt is allowed to soak in a large quantity of water, and the solution thus obtained is kept warm so that it may ferment. This charges the solution with gas and at the same time produces other substances which are supposed to give the bread a good flavour. Such a solution too retains gas much better than pure water. This solution is then mixed with a proper proportion of flour inside a closed vessel, carbon dioxide gas made by the action of acid on a carbonate being pumped into the vessel whilst the mixing is in progress. The mixing is of course effected by mechanical means. As soon as the dough is sufficiently mixed, it is allowed to escape by opening a large tap at the bottom of the mixing vessel. This it does quite readily being forced out by the pressure of gas inside. As it comes out portions of suitable size to make a loaf are cut off. These are at once moulded into loaves and put into the oven. The gas which they contain expands, and light well risen bread is produced. This process is especially suited for wholemeal and other flours containing much offal, which apparently do not give the best results when submitted to the ordinary yeast fermentation.

Before closing this chapter it may be of interest to add a short account of the sale of bread. Bread is at the present time nominally sold by weight under acts of Parliament passed about 80 years ago. That is to say, a seller of bread must provide in his shop scales and weights which will enable him to weigh the loaves he sells. No doubt he would be prepared to do so if requested by a customer, in which case he would probably make up any deficiency in weight which might be found by adding as a makeweight a slice from another loaf. For this purpose it is commonly accepted that the ordinary loaf should weigh two pounds. But in practice this does not occur, for practically the whole of the bread which is sold in this country is sold from the baker's cart, which delivers bread at the houses of customers, and not over the counter. Customers obviously cannot be expected to wait at their doors whilst the man in the cart weighs each loaf he is delivering to them. In actual practice therefore the bread acts, as they are called, are really a dead letter, and bread is sold by the loaf and not by weight, though it must be remembered that the loaf has the reputed weight of two pounds. There are no doubt slight variations from this weight, but for all practical purposes competition nowadays is quite as effective a check on the *bona fides* of the bread seller as enforced sale by weight would be likely to be. If a baker got the reputation of selling loaves appreciably under weight his custom would very soon be transferred to one of his more scrupulous competitors. Altogether it may be concluded that the present unregulated method of sale does not work to the serious disadvantage of the consumers. A little consideration will show that the sale of bread could only be put on a more scientific basis by the exercise of an enormous amount of trouble, and the employment of a very numerous and expensive staff. No doubt the ideally perfect way of regulating the sale of either bread or any other feeding stuff would be to enact that it should be sold by weight, and that the seller should be compelled to state the percentage composition, so that the buyer could calculate the price he was asked to pay per unit of actual foodstuff. Now bread normally contains 36 per cent. of water, but this amount varies greatly. A two pound loaf kept in a dry place may easily lose water by evaporation at the rate of more than an ounce a day. The baker usually weighs out 2 lbs. 3 ozs. of dough to make each two pound loaf, and this amount yields a loaf which weighs in most cases fully two pounds soon after it comes out of the oven. But if the weather is hot and dry such a loaf may very well weigh less than two pounds by the time it is delivered to the consumer. In other words the baker cannot have the weight of the loaves he sells under complete control. Furthermore the loss in weight when a loaf gets dry by evaporation is due entirely to loss of water, and does not decrease the amount of actual foodstuff in the loaf. To sell bread in loaves guaranteed to contain a definite weight of actual foodstuff might be justified scientifically, but practically it would entail so great an expense for the salaries of the inspectors and analysts required to enforce such a regulation that the idea is quite out of the question. Practically, therefore, the situation is that it would be unfair to enforce sale by weight pure and simple for the weight of a loaf varies according to circumstances which are outside the baker's control, and further because the weight of the loaf is no guarantee of the weight of foodstuff present in it. Nor is it possible to enforce sale by guarantee of the weight of foodstuff in the loaf, for to do so would be too troublesome and expensive. Finally the keenness of competition in the baking trade may be relied on to keep an efficient check on the interests of the consumer. Quite recently an important public authority has published the results of weighing several thousand loaves of bread purchased within its area of administration. The results show that over half the two pound loaves purchased were under weight to the extent of five per cent. on the average. Legislation is understood to be suggested as the result of this report, in which case it is to be hoped that account will be taken of the fact that the food value of a loaf depends not only on its weight but also on the percentage of foodstuffs and water which it contains.

CHAPTER VII

THE COMPOSITION OF BREAD

BREAD is a substance which is made in so many ways that it is quite useless to attempt to give average figures showing its composition. It will suffice for the present to assume a certain composition which is probably not far from the truth. This will serve for a basis on which to discuss certain generalities as to the food-value of bread. The causes which produce variation in composition will be discussed later, together with their effect on the food value as far as information is available. The following table shows approximately the composition of ordinary white bread as purchased by most of the population of this country.

VII] THE COMPOSITION OF BREAD

		per cent.
Water		36
Organic substances:		
Proteins	10	
Starch	42	
Sugar, etc.	10	
Fat	1	
Fibre	·3	63·3
Ash:		
Phosphoric acid	·2	
Lime, etc.	·5	·7
		100·0

The above table shows that one of the most abundant constituents of ordinary bread is water. Flour as commonly used for baking, although it may look and feel quite dry, is by no means free from water. It holds on the average about one-seventh of its own weight or 14 per cent. In addition to this rather over one-third of its weight of water or about 35 to 40 per cent. is commonly required to convert ordinary flour into dough. It follows from this that dough will contain when first it is mixed somewhere about one-half its weight of water or 50 per cent. About four per cent. of the weight of the dough is lost in the form of water by evaporation during the fermentation of the dough before it is scaled and moulded. Usually 2 lb. 3 oz. of dough will make a two pound loaf, so that about three ounces of water are evaporated in the oven. This is about one-tenth the weight of the dough or 10 per cent. Together with the four per cent. loss by evaporation during the fermenting period, this makes a loss of water of about 14 per cent., which, when subtracted from the 50 per cent. originally present in the dough, leaves about 36 per cent. of water in the bread. As pointed out in the previous chapter this quantity is by no means constant even in the same loaf. It varies from hour to hour, falling rapidly if the loaf is kept in a dry place.

To turn now to the organic constituents. The most important of these from the point of view of quantity is starch, in fact this is the most abundant constituent of ordinary bread. Nor is it in bread only that starch is abundant. It occurs to the extent of from 50 to 70 per cent. in all the cereals, grains, wheat, barley, oats, maize, and rice. Potatoes too contain about 20 per cent. of starch, in fact it is present in most plants. Starch is a white substance which does not dissolve in cold water, but when boiled in water swells up and makes a paste, which becomes thick and semisolid on cooling. It is this property which makes starch valuable in the laundry. Starch is composed of the chemical elements carbon, hydrogen, and oxygen. When heated in the air it will burn and give out heat, but it does not do so as readily as does fat or oil. It is this property of burning and giving out heat which makes starch

VII] THE COMPOSITION OF BREAD 111

valuable as a foodstuff. When eaten in the form of bread, or other article of food, it is first transformed by the digestive juices of the mouth and intestine into sugar, which is then absorbed from the intestine into the blood, and thus distributed to the working parts of the body. Here it is oxidized, not with the visible flame which is usually associated with burning, but gradually and slowly, and with the formation of heat. Some of this heat is required to keep up the temperature of the body. The rest is available for providing the energy necessary to carry on the movements required to keep the body alive and in health. Practically speaking therefore starch in the diet plays the same part as fuel in the steam engine. The food value of starch can in fact be measured in terms of the quantity of heat which a known weight of it can give out on burning. This is done by burning a small pellet of starch in a bomb of compressed oxygen immersed in a measured volume of water. By means of a delicate thermometer the rise of temperature of the water is measured, and it is thus found that one kilogram of starch on burning gives out enough heat to warm 4·1 kilograms of water through one degree. The quantity of heat which warms one kilogram of water through one degree is called one unit of heat or calorie, and the amount of heat given out by burning one kilogram of any substance is called its heat of combustion or fuel-value. Thus the heat of combustion or fuel-value of starch is 4·1 calories.

Sugar has much the same food-value as starch, in fact starch is readily changed into sugar by the digestive juices of the alimentary canal or by the ferments formed in germinating seeds. From the point of view of food-value sugar may be regarded as digested starch. Like starch, sugar is composed of the elements carbon, hydrogen, and oxygen. Like starch too its value in nutrition is determined by the amount of heat it can give out on burning, and again its heat of combustion or fuel value 3·9 calories is almost the same as that of starch. It will be noted that the whole of the 10 per cent. quoted in the table as sugar, etc., is not sugar. Some of it is a substance called dextrin which is formed from starch by the excessive heat which falls on the outside of the loaf in the oven. Starch is readily converted by heat into dextrin, and this fact is applied in many technical processes. For instance much of the gum used in the arts is made by heating starch. The outside of the loaf in the oven gets hot enough for some of the starch to be converted into dextrin. Dextrin is soluble in water like sugar and so appears with sugar in the analyses of bread. From the point of view of food-value this is of no consequence, as dextrin and sugar serve the same purpose in nutrition, and have almost the same value as each other and as starch.

THE COMPOSITION OF BREAD

Bread always contains a little fat, not as a rule more that one or two per cent. But although the quantity is small it cannot be neglected from the dietetic point of view. Fat is composed of the same elements as starch, dextrin, and sugar, but in different proportions. It contains far less oxygen than these substances. Consequently it burns much more readily and gives out much more heat in the process. The heat of combustion or fuel value of fat is 9·3 calories or 2·3 times greater than that of starch. Evidently therefore even a small percentage of fat must materially increase the fuel value of any article of food. But fat has an important bearing on the nutritive value of bread from quite another point of view. In the wheat grain the fat is concentrated in the germ, comparatively little being present in the inner portion of the grain. Thus the percentage of fat in any kind of bread is on the whole a very fair indication of the amount of germ which has been left in the flour from which the loaf was made. It is often contended nowadays that the germ contains an unknown constituent which plays an important part in nutrition, quite apart from its fuel-value. On this supposition the presence of much fat in a sample of bread indicates the presence of much germ, and presumably therefore much of this mysterious constituent which is supposed to endow such bread with a special value in the nutrition particularly of young children. This question will be discussed carefully in a later chapter.

White bread contains a very small percentage of what is called by analysts fibre. The quantity of this substance in a food is estimated by the analyst by weighing the residue which remains undigested when a known weight of the food is submitted to a series of chemical processes designed to imitate as closely as may be the action of the various digestive juices of the alimentary canal. Theoretically, therefore, it is intended to represent the amount of indigestible matter present in the food in question. Practically it does not achieve this result for some of it undoubtedly disappears during the passage of the food through the body. It is doubtful however if the portion which disappears has any definite nutritive value. That part of the fibre which escapes digestion and is voided in the excrement cannot possibly contribute to the nutrition of the body. Nevertheless it exerts a certain effect on the well-being of the consumer, for the presence of a certain amount of indigestible material stimulates the lower part of the large intestine and thus conduces to regularity in the excretion of waste matters, a fact of considerable importance in many cases. The amount of fibre is an index of the amount of indigestible matter in a food. In white bread it is small. In brown breads which contain considerable quantities

VII] THE COMPOSITION OF BREAD

of the husk of the wheat grain it may be present to the extent of two or three per cent. Such breads therefore will contain much indigestible matter, but they will possess laxative properties which make them valuable in some cases.

We have left to the last the two constituents which at the present time possess perhaps the greatest interest and importance, the proteins and the ash. The proteins of bread consist of several substances, the differences between which, for the present purpose, may be neglected, and we may assume that for all practical purposes the proteins of bread consist of one substance only, namely gluten. The importance of gluten in conferring on wheat flour the property of making light spongy loaves has already been insisted upon. No doubt this property of gluten has a certain indirect bearing on the nutritive value of bread by increasing its palatability. But gluten being a protein has a direct and special part to play in nutrition, which is perhaps best illustrated by following one step further the comparison between the animal body and a steam engine. It has been pointed out that starch, sugar, and fat play the same part in the body as does the fuel in a steam engine. But an engine cannot continue running very long on fuel alone. Its working parts require renewing as they wear away, and coal is no use for this purpose. Metal parts must be renewed with metal. In much the same way the working parts of the animal body wear away, and must be renewed with the stuff of which they are made. Now the muscles, nerves, glands and other working parts of the body are made of protein, and they can only be renewed with protein. Consequently protein must be supplied in the diet in amount sufficient to make good from day to day the wear and tear of the working parts of the body. It is for this reason that the protein of bread possesses special interest and importance.

Protein like starch, sugar, and fat contains the elements carbon, hydrogen, and oxygen, but it differs from them in containing also a large proportion of the element nitrogen, which may be regarded as its characteristic constituent. When digested in the stomach and intestine it is split into a large number of simpler substances known by chemists under the name of amino-acids. Every animal requires these amino-acids in certain proportions. From the mixture resulting from the digestion of the proteins in its diet the amino-acids are absorbed and utilised by the body in the proportions required. If the proteins of the diet do not supply the amino-acids in these proportions, it is obvious that an excessive amount of protein must be provided in order that the diet may supply enough of that particular amino-acid which is present in deficient amount, and much of those amino-acids which are abundantly present must

VII] THE COMPOSITION OF BREAD 117

go to waste. This is undesirable for two reasons. Waste amino-acids are excreted through the kidneys, and excessive waste throws excessive work on these organs, which may lead to defective excretion, and thus cause one or other of the numerous forms of ill health which are associated with this condition. Again, excessive consumption of protein greatly adds to the cost of the diet, for protein costs nearly as many shillings per pound as starch or sugar costs pence.

These considerations show clearly the wisdom of limiting the amount of protein in the diet to the smallest amount which will provide for wear and tear of the working parts. The obvious way to do this is to take a mixed diet so arranged that the various articles of which the diet consists contain proteins which are so to speak complementary. The meaning of this is perhaps best illustrated by a concrete example. The protein of wheat, gluten, is a peculiar one. On digestion it splits like other proteins into amino-acids, but these are not present from the dietetic point of view in well balanced proportions. One particular amino-acid, called glutaminic acid, preponderates, and unfortunately this particular acid does not happen to be one which the animal organism requires in considerable quantity. Other amino-acids which the animal organism does require in large amounts are deficient in the mixture of amino-acids yielded by the digestion of the protein of wheat. It follows, therefore, that to obtain enough of these latter acids a man feeding only on wheat products would have to eat a quantity of bread which would supply a great excess of the more abundant glutaminic acid, which would go to waste with the evil results already outlined. From this point of view it appears that bread should not form more than a certain proportion of the diet, and that the rest of the diet should consist of foods which contain proteins yielding on digestion little glutaminic acid and much of the other amino-acids in which the protein of wheat is deficient. Unfortunately information as to the exact amount of the different amino-acids yielded by the digestion of the proteins even of many of the common articles of food is not available. But many workers are investigating these matters, and the next great advance in the science of dietetics will probably come along these lines. By almost universal custom certain articles of food are commonly eaten in association: bread and cheese, eggs and bacon, are instances. Such customs are usually found to be based on some underlying principle. The principle in this case may well be that of complementary proteins.

The remarks which have been made above on the subject of the *rôle* of protein in the animal economy apply to adults in which protein is required for wear and tear only and not for increase in weight. They will obviously apply with greatly increased force to the case of growing children, who require protein not only for wear and tear, but for the building up of their muscles and other working parts as they grow and develope. Consequently the diet of children should contain more protein in proportion to their size than that of adults. For this reason it is not desirable that bread should form an excessive proportion of their diet. The bread they eat should be supplemented with some other food richer in protein.

The ash of bread although so small in amount cannot be ignored, in fact it is regarded as of very great importance by modern students of dietetics. The particular constituent of the ash to which most importance is attached is phosphoric acid. This substance is a necessary constituent of the bones and of the brain and nerves of all animals. It exists too in smaller proportions in other organs. Like other working parts of the body the bones and the nervous system are subject to wear and tear, which must be replaced if the body is to remain in normal health. A certain daily supply of phosphoric acid is required for this purpose, and proportionally to their size more for children than for adults. Considerable difference of opinion as to the exact amount required is expressed by those who have investigated this question, nor is it even agreed whether all forms of phosphoric acid are of the same value. There is however a general recognition of the importance of this constituent of the diet, and the subject is under investigation in many quarters.

CHAPTER VIII

CONCERNING DIFFERENT KINDS OF BREAD

The table given in the last chapter states the average composition of ordinary white bread baked in the form of cottage loaves, and the remarks on the various constituents of bread in the preceding pages have for the most part referred to the same material, though many of them may be taken to refer to bread in general. It will now be of interest to inquire as to the variation in composition which is found among the different kinds of bread commonly used in this country. This enquiry will be most readily conducted by first considering the possible causes which may affect the composition of bread.

The variation in the composition of bread is a subject which is taken up from time to time by the public press, and debated therein with a great display of interest and some intelligent knowledge. In most of the press discussions in the past interest has been focussed almost entirely on the effect of different kinds of milling. The attitude commonly assumed

VIII] DIFFERENT KINDS OF BREAD

by the food reform section of the contributors may be stated shortly as follows: In the days of stone milling a less perfect separation of flour and bran was effected, and the flour contained more of the materials situated in the grain near the husk than do the white flours produced by modern methods of roller milling. Again the modern roller mills separate the germ from the flour, which the stone mills fail to do, at any rate so completely. Thus the stone ground flours contain about 80 per cent. of the grain, whilst the whole of the flour obtained from the modern roller mill seldom amounts to much more than about 72 per cent. The extra eight per cent. of flour produced in the stone mills contains all or nearly all the germ and much of the material rich in protein which lies immediately under the husk. Hence the stone ground flour is richer in protein, and in certain constituents of the germ, than white roller mill flour, and hence again stone ground flour has a higher nutritive value. Roller mill flour has nothing to commend it beyond its whiteness. It has been suggested that millers should adopt the standard custom of producing 80 per cent. of flour from all the wheat passing through their mills and thus retain those constituents of the grain which possess specially great nutritive value.

It would probably be extremely difficult to produce 80 per cent. of flour from many kinds of wheat, but for the present this point may be ignored, whilst we discuss the variation in the actual chemical composition of the flour produced as at present and on the 80 per cent. basis. In comparing the chemical composition of different kinds of flour it is obvious that the flours compared must have been made from the same lot of wheat, for as will be seen later different wheats vary greatly in the proportions of protein and other important constituents which they contain. Unfortunately the number of analyses of different flours made from the same lots of wheat is small. Perhaps the best series is that published by Dr Hamill in a recent report of the Local Government Board. Dr Hamill gives the analyses of five different grades of flour made at seven mills, each mill using the same blend of wheats for all the different kinds of flour. Calculating all these analyses to a basis of 10 per cent. of protein in the grade of flour

Description of flour or offal	Protein per cent.	Phosphoric acid per cent.
Flours:		
Patents	10.0	0.18
Straight grade, about 70 per cent.	10.6	0.21
Households	10.9	0.26
Standard flour, about 80 per cent.	11.0	0.35
Wholemeal	11.3	0.73
Offals:		
Germ	24.0	2.22
Sharps	14.5	1.66
Bran	13.5	2.50

VIII] DIFFERENT KINDS OF BREAD 123

known as patents, the figures on the opposite page were obtained, which may be taken to represent with considerable accuracy the average composition of the various kinds of flours and offals when made from the same wheat.

Accepting these figures as showing the relative proportions of protein and phosphoric acid in different flours as affected by milling only, other sources of variation having been eliminated by the use of the same blend of wheat, it appears that the flours of commercially higher grade undoubtedly do contain somewhat less protein and phosphoric acid than lower grade or wholemeal flours. Taking the extreme cases of patents and wholemeal flours, the latter contains one-ninth more protein and four times more phosphoric acid than the former, provided both are derived from the same wheat.

In actual practice, however, it generally happens that the higher grade flours are made from a blend of wheats containing a considerable proportion of hard foreign wheats which are rich in nitrogen, whilst wholemeal and standard flours are usually made from home grown wheats which are relatively poor in nitrogen. From a number of analyses of foreign and home grown wheats it appears that the relative proportions of protein is about $12\frac{1}{2}$ per cent. in the hard foreign wheats as compared with 10 per cent. in home grown wheats. Thus the presence of a larger proportion of protein in the hard wheats used in the blend of wheat for making the higher grade flours must tend to reduce the difference in protein content between say patents and wholemeal flours as met with in ordinary practice. Furthermore much of the bread consumed by that part of the population to whom a few grams per day of protein is of real importance is, or should be, made, for reasons of economy, from households flour, and the disparity between this grade of flour and wholemeal flour is much less than is the case with patents. It appears, therefore, on examining the facts, that there is no appreciable difference in the protein content of the ordinary white flours consumed by the poorer classes of the people and wholemeal flour or standard flour.

In the above paragraphs account has been taken only of the total amount of protein in the various kinds of bread and flour. It is obvious, however, that the total amount present is not the real index of food-value. Only that portion of any article of diet which is digested in the alimentary canal can be absorbed into the blood and carried thereby to the tissues where it is required to make good wear and tear. The real food-value must therefore depend not on the total amount of foodstuff present but on the amount which is digestible. The proportion of protein which can be digested in the different kinds

VIII] DIFFERENT KINDS OF BREAD 125

of bread has been the subject of careful experiments in America, and lately in Cambridge. The method of experimenting is arduous and unpleasant. Several people must exist for a number of days on a diet consisting chiefly of the kind of bread under investigation, supplemented only by small quantities of food which are wholly digestible, such as milk, sugar and butter. During the experimental period the diet is weighed and its protein content estimated by analysis. The excreta are also collected and their protein content estimated by analysis, so that the amount of protein which escapes digestion can be ascertained. The experiment is then repeated with the same individuals and the same conditions in every way except that another kind of bread is substituted for the one used before. From the total amount of protein consumed in each kind of bread the total amount of protein voided in the excreta is subtracted, and the difference gives the amount which has been digested and presumably utilised in the body. From these figures it is easy to calculate the number of parts of protein digested for every 100 parts of protein eaten in each kind of bread. This description will have made evident the unpleasant nature of such experimental work. Its laboriousness will be understood from the fact that a series of experiments of this kind carried out at Cambridge last winter necessitated four people existing for a month on the

126 THE STORY OF A LOAF OF BREAD [CH.

meagre diet above mentioned, and entailed over 1000 chemical analyses.

The following table shows the amounts of protein digested per 100 parts of protein consumed in bread made from various kinds of flour, as based on the average of a number of experiments made in America, and in the experiments at Cambridge above referred to.

Kind of flour from which bread was made	Percentage of the grain contained in the flour	Amount of protein digested per 100 parts eaten	
		American experiments	Cambridge experiments
Patents	36	—	89
Straight grade	70	89	—
Standard	80	81	86
Brown	88	—	80
Brown	92	—	77
Wholemeal	100	76	—

The American and the Cambridge figures agree very well with each other, and this gives confidence in the reliability of the results. It appears to be quite certain therefore that the protein in bread made from the higher grade flours is very considerably more digestible than that contained in bread made from flours containing greater amounts of husk. The percentages following the names of the various grades of flour in the first column of the table indicate approximately the proportion of the whole grain which went into the flour to which the figure is

VIII] DIFFERENT KINDS OF BREAD 127

attached. Looking down these figures it appears that the digestibility of the protein decreases as more and more of the grain is included in the flour. It follows, therefore, that whilst by leaving more and more of the grain in the flour we increase the percentage of protein in the flour, and consequently in the bread, at the same time we decrease the digestibility of the protein. Apparently, too, this decrease in digestibility is proportionally greater than the increase in protein content, and it follows therefore that breads made from low grade flours containing much husk will supply less protein which is available for the use of the body, although they may actually contain slightly more total protein than the flours of higher grade.

When all the facts are taken into account it appears that the contention of the food reformers, that the various breads which contain those constituents of the grain which lie near the husk are capable of supplying more protein for the needs of the body than white breads, cannot be upheld. From statistics collected by the Board of Trade some few years ago as to the dietary of the working classes it appears that the diet of workers both in urban and in rural districts contains about 97 grams of total protein per head per day. This is rather under than over the commonly accepted standard of 100 grams of protein which is supposed to be required daily by a healthy man at moderate work. Consequently a change in his diet which increased the amount of protein might be expected to be a good change. But the suggested change of brown bread for white, though it appears to increase the total protein, turns out on careful examination to fail in its object, for it does not increase the amount of protein which can be digested.

From the same statistics it appears that the diet of a working man includes on the average about $1\frac{1}{4}$ lb. of bread per day. This amount of bread contains about 60 grams of protein, or two-thirds of the total protein of the diet. Now it was pointed out in the last chapter that the protein of wheat was very rich in glutaminic acid, a constituent of which animals require comparatively small amounts. It is also correspondingly poor in certain constituents which are necessary to animals. Apparently therefore it would be better to increase the diet in such cases by adding some constituent not made from wheat than by changing the kind of bread. From the protein point of view, however we look at it, there appears to be no real reason for substituting one or other of the various kinds of brown bread for the white bread which seems to meet the taste of the present day public.

But important as protein is it is not everything in a diet. As we have already pointed out the food

VIII] DIFFERENT KINDS OF BREAD 129

must not only repair the tissues, it must also supply fuel. It has been shown also that the fuel-value of a food can be ascertained by burning a known weight and measuring the number of units of heat or calories produced. Many samples of bread have been examined in this way in the laboratories of the American Department of Agriculture, and it appears from the figures given in their bulletins that the average fuel value of white bread is about 1·250 calories per pound, of wholemeal bread only 1·150 calories per pound. These figures are quite in accord with those which were obtained in Cambridge in 1911, in connection with the digestion experiments already described, which were also extended so as to include a determination of the proportion of the energy of the bread which the diet supplied to the body. The energy or fuel-value of the diet was determined by measuring the amount of heat given out by burning a known weight of each of the kinds of bread used in the experiment. The energy which was not utilised by the body was then determined by measuring how much heat was given out by burning the excreta corresponding to each kind of bread. The following table gives side by side the average results obtained in several such experiments in America and in Cambridge.

The agreement between the two sets of figures is again on this point quite satisfactory. It is evident

130 THE STORY OF A LOAF OF BREAD [CH.

Kind of flour from which the bread was made	Percentage of the grain contained in the flour	Amount of energy utilised per 100 units in food	
		American experiments	Cambridge experiments
Patents	36	96	96
Straight grade	70	92	—
Standard	80	87	95
Brown	88	—	90
Brown	92	—	89
Wholemeal	100	82	—

that a greater proportion of the total energy of white bread can be utilised by the body than is the case with any of the breads made from flours of lower commercial grades which contain more husk. In fact it appears that the more of the outer parts of the grain are left in the flour the smaller is the proportion of the total energy of the bread which can be utilised. Combining this conclusion with the fact that brown breads contain on the average less total energy than white breads, there can be no doubt that white bread is considerably better than any form of brown bread as a source of energy for the body.

There is one more important substance in respect of which great superiority is claimed for brown breads, namely phosphoric acid. From the table on page 122 there can be no doubt that flours containing more of the outer parts of the grain are very much richer in phosphoric acid than white flours, and the disparity is so great that after allowing for the larger proportion of water in brown breads they must contain far more of this substance than do white breads. In the Cambridge digestibility experiments quoted above the proportion of the phosphoric acid digested from the different breads was determined. It was found that for every 100 parts of phosphoric acid in white bread only 52 parts were digested, and that in the case of the brown breads this proportion fell to 41 parts out of 100. Again, as in the case of protein and energy, the phosphoric acid in white bread is more readily available to the body than that of brown bread, but in this case the difference in digestibility is not nearly enough to counterbalance the much larger proportion of phosphoric acid in the brown bread. There is no doubt that the body gets more phosphoric acid from brown bread than from the same quantity of white bread. But before coming to any practical conclusion it is necessary to know two things, how much phosphoric acid does a healthy man require per day, and does his ordinary diet supply enough?

From the Board of Trade statistics already quoted it appears that, on the assumption that the average worker eats white bread only, his average diet contains 2·4 grams of phosphoric acid per day, which would be raised to 3·2 grams if the white bread were replaced by bread made from 80 per cent. flour containing ·35 per cent. of phosphoric acid. Information as to the amount of phosphoric acid required per day by a healthy man is somewhat scanty, and indicates that the amount is very variable, but averages about $2\frac{1}{2}$ grams per day. If this is so, the ordinary diet with white bread provides on the average enough phosphoric acid. Exceptional individuals may, however, be benefited by the substitution of brown bread for white, but it would probably be better even in such cases, for the reasons stated when discussing the protein question, to raise the phosphorus content of their diet by the addition of some substance rich in phosphorus but not made from wheat.

Finally comes the question of the variation in the composition of bread due to the presence or absence of the germ. The first point in this connection is to decide whether germ is present in appreciable proportions in any flour except wholemeal. The germ is a soft moist substance which flattens much more readily than it grinds. Consequently it is removed from flour by almost any kind of separation, even when very coarse sieves are employed. If this contention is correct no flour except wholemeal should contain any appreciable quantity of germ, and it is certainly very difficult to demonstrate the presence of actual germ particles even in 80 per cent. flour. Indirect evidence of the presence of germ may, however, be obtained as already explained by estimating

VIII] DIFFERENT KINDS OF BREAD 133

by chemical analysis the proportion of fat present in various flours. The figures for such estimations are given by Dr Hamill in the report of the Local Government Board already referred to. They show that the percentages of fat in different grades of flours made from the same blends of wheat are on the average of seven experiments as follows : patents flour ·96: household flours 1·25: 80 per cent. or standard flour 1·42. These figures show that the coarser flours containing more of the whole grain do contain more germ than the flours of commercially higher grade, in spite of the fact that it is difficult to demonstrate its presence under the microscope.

Remembering, however, that the whole of the germ only amounts to about $1\frac{1}{2}$ per cent. of the grain, it is clear that the presence or absence of more or less germ cannot appreciably affect the food-value as measured by protein content or energy-value. It is still open to contention that the germ may contain some unknown constituent possessing a peculiar effect on nutrition. Such a state of things can well be imagined in the light of certain experimental results which have been obtained during the last few years.

It has been shown for instance by Dr Hopkins in Cambridge, and his results have been confirmed at the Carnegie Institute in America, that young rats fail to thrive on a diet composed of suitable amounts of purified protein, fat, starch, and ash, but that they thrive and grow normally on such a diet if there is added a trace of milk or other fresh animal or vegetable substance far too small to influence either the protein content or the energy-value. Another case in point is the discovery that the disease known as beri beri, which is caused by a diet consisting almost exclusively of rice from which the husk has been removed, can be cured almost at once by the administration of very small doses of a constituent existing in minute quantities in rice husk. The suggestion is that high grade flours, like polished rice, may fail to provide some substance which is necessary for healthy growth, a substance which is removed in the germ or husk when such flours are purified, and which is present in flours which have not been submitted to excessive purification.

The answer is that no class in Great Britain lives on bread exclusively. Bread appears from the government statistics already quoted to form only about half the diet of the workers of the country. Their diet includes also some milk, meat, and vegetables, and such substances, according to Dr Hopkins' experiments, certainly contain the substance, whatever it may be, that is missing from the artificial diet on which his young rats failed to thrive.

One last point. It will have been noticed in the figures given above that the variations in protein content, digestibility, and energy-value, between different kinds of bread are usually not very large. There is, however, one constituent of all breads whose proportions vary far more widely, namely water. During last summer the author purchased many samples of bread in and around Cambridge, and determined the percentage of water in each sample. The samples were all one day old so that they are comparable with one another The results on the whole are a little low, probably because the work was done during a spell of rather dry weather, when the loaves would lose water rapidly.

The average figures are summarised below :

	Percentage of water
Cottage loaves made of white flour	31·7
Tinned ,, ,, ...	32·7
Small fancy loaves made of white flour ...	33·7
Tinned loaves made of "Standard" flour ...	35·9
Tinned ,, brown or germ flour ...	40·0

The figures speak for themselves. There must obviously be more actual food in a cottage loaf of white flour containing under 32 per cent. of water than in any kind of Standard or brown loaf in which the percentage of water is 36 to 40. It is quite extraordinary that no one who has organised any of the numerous bread campaigns in the press appears to have laid hold of the enormous variation in the water content of different kinds of bread, and its obvious bearing on their food-value.

BIBLIOGRAPHY

THE reader who wishes further information on any of the numerous subjects connected with the growth, manipulation and composition of breadstuffs is referred to the following publications, to which among others the author is much indebted. The list is arranged, as far as possible, in the same order as the chapters of the book.

CHAPTER I.

The Book of the Rothamsted Experiments, by A. D. Hall. (John Murray, 1905.)

The Feeding of Crops and Stock, by A. D. Hall. (John Murray, 1911.)

Fertilizers and Manures, by A. D. Hall. (John Murray, 1909.)

The Soil, by A. D. Hall. (John Murray, 1908.)

Agriculture and Soils of Kent, Surrey, and Sussex, by A. D. Hall and E. J. Russell. (Board of Agriculture and Fisheries.)

Some Characteristics of the Western Prairie Soils of Canada, by Frank T. Shutt. (*Journal of Agricultural Science*, Vol. III, p. 335.)

Dry Farming: its Principles and Practice, by Wm Macdonald. (T. Werner Laurie.)

Profitable Clay Farming, by John Prout. (1881.)

Continuous Corn Growing, by W. A. Prout and J. Augustus Voelcker. (*Journal of the Royal Agricultural Society of England*, 1905.)

CHAPTER II.

The Wheat Problem, by Sir W. Crookes. (John Murray, 1899.)

The Production of Wheat in the British Empire, by A. E. Humphries. (*Journal of the Royal Society of Arts*, Vol. LVII, p. 229.)

Wheat Growing in Canada, the United States, and the Argentine, by W. P. Rutter. (Adam and Charles Black, 1911.)

Agricultural Note-Book, by Primrose McConnell. (Crosby, Lockwood and Son, 1910.)

BIBLIOGRAPHY

CHAPTER III.

Agricultural Botany, by J. Percival. (Duckworth and Co., 1900.)

The Interpretation of the Results of Agricultural Experiments, by T. B. Wood, and Field Trials and their interpretation, by A. D. Hall and E. J. Russell. (*Journal of the Board of Agriculture and Fisheries*, Supplement No. 7, Nov. 1911.)

Heredity in Plants and Animals, by T. B. Wood and R. C. Punnett. (*Journal of the Highland and Agricultural Society of Scotland*, Vol. xx, Fifth Series, 1908.)

Mendelism, by R. C. Punnett. (Macmillan and Co., 1911.)

Mendel's Laws and Wheat Breeding, by R. H. Biffen. (*Journal of Agricultural Science*, Vol. I, p. 4.)

Studies in the Inheritance of Disease Resistance, by R. H. Biffen. (*Journal of Agricultural Science*, Vol. II, p. 109; Vol. IV, p. 421.)

The Inheritance of Strength in Wheat, by R. H. Biffen. (*Journal of Agricultural Science*, Vol. III, p. 86.)

Variation, Heredity, and Evolution, by R. H. Lock. (John Murray, 1909.)

Minnesota Wheat Breeding, by Willet M. Hays and Andrew Boss. (McGill-Warner Co., St Paul.)

The Improvement of English Wheat, by A. E. Humphries and R. H. Biffen. (*Journal of Agricultural Science*, Vol. II, p. 1.)

Plant Breeding in Scandinavia, by L. H. Newman. (The Canadian Seed Growers Association, Ottawa, 1912.)

CHAPTERS IV, V, AND VI.

The Technology of Bread Making, by W. Jago. (Simpkin, Marshall and Co., 1911.)

Modern Development of Flour Milling, by A. E. Humphries. (*Journal of the Royal Society of Arts*, Vol. LV, p. 109.)

Home Grown Wheat Committee's Reports. (59, Mark Lane, London, E.C.)

The Chemistry of Strength of Wheat Flour, by T. B. Wood. (*Journal of Agricultural Science*, Vol. II, pp. 139, 267.)

BIBLIOGRAPHY

CHAPTERS VII AND VIII.

Composition and Food Value of Bread, by T. B. Wood. (*Journal of the Royal Agricultural Society of England*, 1911.)

Some Experiments on the Relative Digestibility of White and Wholemeal Breads, by L. F. Newman, G. W. Robinson, E. T. Halnan, and H. A. D. Neville. (*Journal of Hygiene*, Vol. XII, No. 2.)

Nutritive Value of Bread, by J. M. Hamill. (*Local Government Board Report*, Cd. 5831.)

Bleaching and Improving Flour, by J. M. Hamill and G. W. Monier Williams. (*Local Government Board Report*, Cd. 5613.)

Diet of Rural and Urban Workers. (*Board of Trade Reports*, Cd. 1761 and 2337.)

Bulletins of the U.S.A. Department of Agriculture. (Division of Chemistry 13; Office of Experiment Stations 21, 52, 67, 85, 101, 126, 156, 185, 227.)

INDEX

Aerated bread, 104
Amino-acids, 116
Ash of bread, 119

Baking, 63, 91
Baking powders, 102
Biffen's new varieties, 49, 59
  method, 41, 46, 58
Bread, amount in diet, 127
  composition of, 109
  variations in, 120
  water in, 135
Break rolls, 81
Breeding of wheat, 29, 35, 40
Burgoyne's Fife, 59

Climate suitable for wheat, 2, 28
Clover as preparation for wheat, 8
Colloids, 67
Continuous growth of wheat, 7
Crookes, Sir W., shortage of nitrogen, 5
Cropping power of wheats, 32
Cross-breeding, 40

Digestibility of bread, 124
Dressing wheat, 16
Dry farming, 10

Elements required by wheat, 2
Energy-values, 111

Fat in bread, 113
Fermentation in dough, 94
Fibre, 114

Field plots, accuracy of, 32
Fife wheat, 47, 57
Flour, composition of, 122
  grades of, 88, 122
  self-rising, 102
Food-value of various breads, 120
  of starch, etc. in bread, 110
Foreign wheat growing, 21
Fuel-values, 111
Futures, 26

Germ, food-value of, 132
  in milling, 89
  in bread, 114, 132
Gluten, 63
  properties of, 68
Grades of flour, 88, 122
  of wheat, 23

Home Grown Wheat Committee, 53, 56
Hopkins' work, 132
Hybridisation, 40

Johannsen, 37
Judging wheats, 60

Improvers, flour, 100
Indigestible matter in bread, 114
Inheritance in wheat, 41

Lawes and Gilbert, 4
Liebig, 3
Little Joss wheat, 51

INDEX

Manuring wheat, 3, 7
Markets, home, 16
  foreign, 22, 27
Market quotations, 19
Mendel's laws, 40
Milling, history of, 74, 77
  effect of, on flour, 122
Mineral manures, 3
Minnesota experiments, 36

Natural moisture in wheat, 52
Nitrogen, cost of, in manures, 4
  fixation, 8
  for wheat, 4
  from air, 6
  scarcity of, 5
  synthetic, 6

Ovens for baking, 99

Patents flour, 88
Pedigree in wheat, 39
Phosphates in bread, 119
  in diet, 131
  in flour, 70
Plots for yield testing, 32
Protein, cost of, in diet, 116
  in bread, 115
Prout's system of farming, 7
Pure-line theory, 37
Purification of flour, 86

Rainfall for wheat, 2
Red Fife, 47
Reduction rolls, 87
Roller mill, 79
Rotation of crops, 9
Rothamsted experiments, 4
Rust-proof wheat, 51

Sale of bread, 105
  of wheat, 16
Scaling loaves, 96
Selection for cropping power, 35
Self-rising flour, 102
Separation of flour, 85
Semolina, 86
Sheep-folding, 10
Soils for wheat, 2
"Standard" flour, 135
Starch in bread, 110
Stone mill, 75
Strength of flour, cause of, 62
  of flour, test for, 66, 72
  of wheat or flour, 53
Strong wheats, characters of, 60
  value of, 59
Sugar in bread, 112
Synthetic nitrogenous manures, 6

Thrashing wheat, 15
Turbidity test for strong wheats, 73

Variety of wheat, choice of, 28
  testing, 32
Virgin soils, 5

Water in bread, 109, 135
Weak wheats, characters of, 61
Weights and measures, 17

Yeast, growth in dough, 94
Yield of wheat, conditions of, 28

For EU product safety concerns, contact us at Calle de José Abascal, 56–1°,
28003 Madrid, Spain or eugpsr@cambridge.org.

www.ingramcontent.com/pod-product-compliance
Ingram Content Group UK Ltd.
Pitfield, Milton Keynes, MK11 3LW, UK
UKHW040157230326
469255UK00012B/146